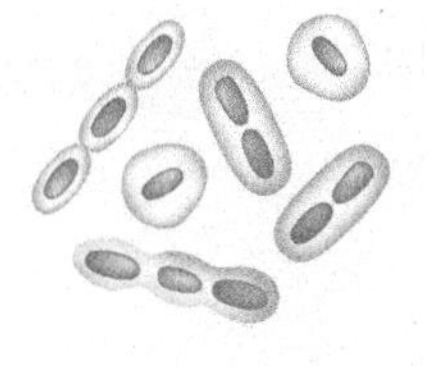

THE GERM CODE

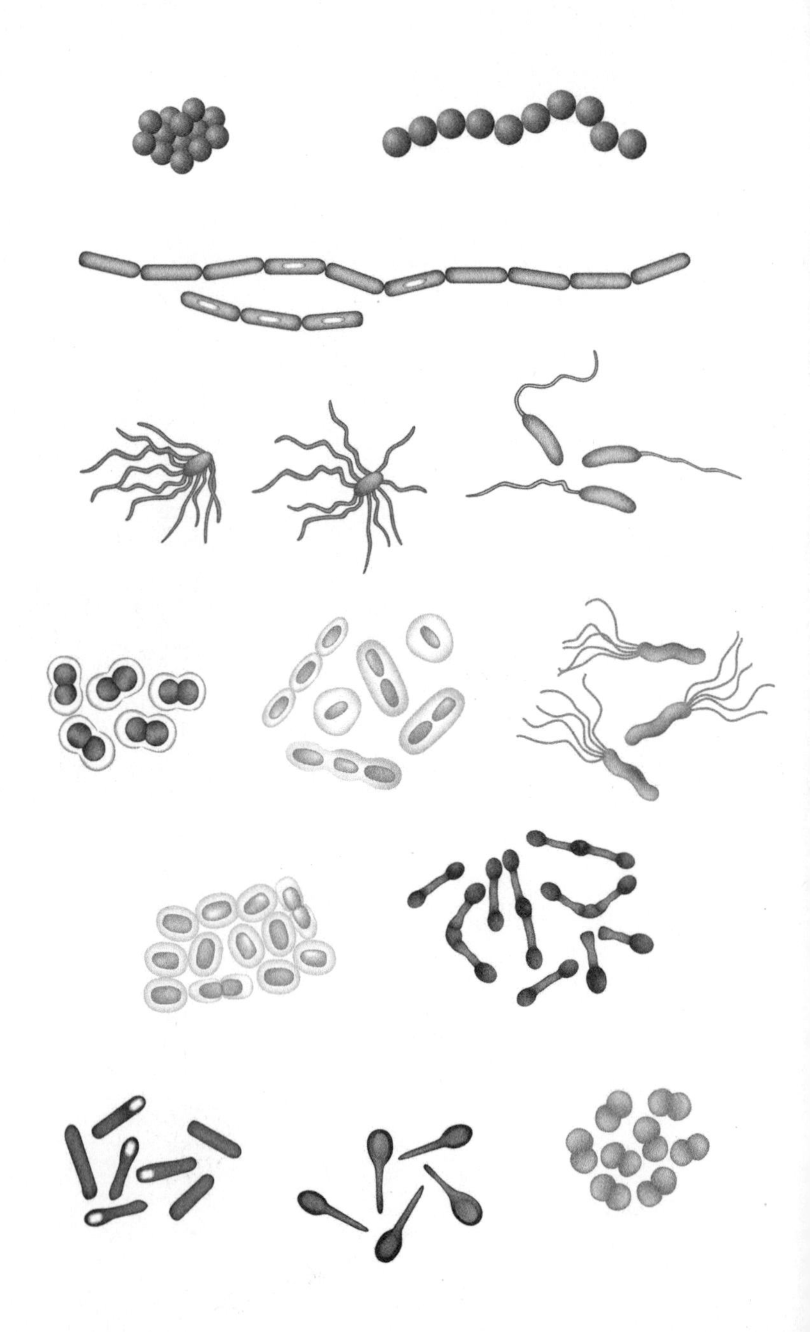

JASON TETRO

THE GERM CODE

HOW TO STOP WORRYING AND LEARN TO LOVE THE MICROBES

Doubleday Canada

Library and Archives of Canada Cataloguing in Publication
is available upon request.

ISBN: 978-0-385-67853-7

The headings for chapters 2 and 12 were replicated from images courtesy of The Centers for Disease Control Prevention: Larry Stauffer and Charles D. Humphrey, respectively.

The heading for chapter 10 was replicated from an image courtesy of Jasper Lawrence [GFDL (http://www.gnu.org/copyleft/fdl.html) or CC-BY-SA-3.0-2.5-2.0-1.0 (http://creativecommons.org/licenses/by-sa/3.0)], via Wikimedia Commons.

The heading for chapter 11 was replicated from an image courtesy of Ude, http://www.tci.uni-hannover.de/ (Own work) [GFDL (http://www.gnu.org/copyleft/fdl.html) or CC-BY-SA-3.0 (http://creativecommons.org/licenses/by-sa/3.0)], via Wikimedia Commons.

Cover and text design: CS Richardson
Illustration on cover, i, ii, 1, 47, 73, and 170:
Alila Medical Images / Shutterstock.com
Printed and bound in the USA

Published in Canada by Doubleday Canada,
a division of Random House of Canada Limited

www.randomhouse.ca

10 9 8 7 6 5 4 3 2 1

To my parents, Peter and Patricia Tetro

CONTENTS

INTRODUCTION

The most important relationship in each of our lives is with germs. They are with us from the moment we are born until our last breath. They surround us, outnumber us, live on and in us, and interact with us, both to our benefit and our detriment. They are essential to keep us healthy, and at times they happen to make us sick and possibly threaten our lives. Yet, despite playing such a pivotal role in our existence, germs are often forgotten or even ignored by us, and we spend little to no time building a lasting bond with them.

Can we be blamed? Not really. Building a successful and balanced rapport between two humans can be difficult enough, never mind a population of unseen entities. For millennia, philosophers, psychologists and communication experts have struggled to explain how we develop and maintain our various connections, both with people and the environment.

Simply defining such a relationship can present a significant challenge. One way to summarize our shared life with germs stems from the work of American author and relationship counsellor John Gray. In the 1990s, he offered what was to become a hugely popular perspective on how humans interrelate. His titular theory, *Men Are from Mars, Women Are from Venus*, has since entered common parlance in discussing the discordance between men and women. As for our existence with germs, a different sort of planetary metaphor might be used to describe how we coexist with microbes: "Humans Are from Earth, Germs Are from Uranus."

The pun is biologically correct, but only up to a point. Germs happen to be found everywhere, and they play an integral role in every aspect of human existence. The sheer ubiquity of germs and humans demands a properly managed coexistence. But since the dawn of humanity, our relationship has been fraught at best. Our history is replete with stories of a love-hate connection that never seems to be balanced. As Gray pointed out in his book, the reason for imbalance stems not from the fact that the two parties are a bad match, but from a mutual lack of understanding of their fundamental differences. As a result, perceptions become lopsided, and rather than developing a coexistence, men and women come into opposition and, eventually, unnecessary war.

In this regard, the balance between humans and germs was doomed from the start, as we didn't realize germs existed until the seventeenth century. We did, however, know the consequences of their presence, manifested in outbreaks, epidemics and pandemics. But before these microscopic creatures were seen under the microscope, their actions were explained in an almost religious way, as if ethereal or otherworldly beings were responsible for the disruption to human existence. Even when the godly or alien context was put to rest, germs were regarded not as entities but as phenomena, such as "miasma," or bad air. Should miasma make an appearance in one's village, death would surely follow.

When germs were eventually recognized as the cause of infection—a momentous occasion some two hundred years after their discovery—we did not marvel at the revelation, nor did we devote the majority of our efforts to understanding the fundamentals of germy life. Instead, we went on the offensive in the hope of ridding the earth of these supposed killers.

This warlike mentality grew over the decades that followed, fostered by negative attitudes towards them in literature, film and other media. While the goal might have been to exploit the public's fear of the unknown, the consequence was our continual inability to think of germs as fellow inhabitants of the earth who deserve respect.

Today, we live in a more enlightened world where our information about germs, while not complete, is still robust enough to give them a second look. We have just as much information on their positive and beneficial contributions to our lives as we do their harmful impact. We have gained an understanding of how they live in the environment and how they interact with us.

They are not always causing infection. Some people have even harnessed germs to help humanity, developing medicinal and nutritional products that improve human health and our environment. Our world has begun to see how achieving coexistence is far more important than the futile attempts at total eradication.

As Gray and so many others have pointed out in their treatises, theses, essays and books, a successful, balanced relationship requires, first, knowledge of the fundamentals that form the basis of a person's actions—otherwise known as the code—and, second, the application of that code in everyday life. The code is usually simplistic—for example, Mars/Venus—while the application requires study and understanding. Consider the most basic code of human relationships, the alphabet. English is one of the languages based on the twenty-six letters of the Latin alphabet. Yet presenting someone who does not speak English with the following:

A B C D E F G H I J K L M N O P Q R S T U V W X Y Z

will do little to provide that person with the ability to communicate. Understanding the *application* of this code, however—including the development of words, the use of grammar and proper pronunciation—will. As anyone who has learned how to speak a language realizes, the knowledge of the code may be fairly easy, but learning how to understand and master its application can be a lengthy, and at times frustrating, experience. Moreover, when presented with the fact that the same code can be applied in a different manner to create entirely different languages, one can see that understanding a code is nothing without knowing how it is applied.

If asked what the germ code might be, most people would probably say something like "They make us sick and can kill us." While not entirely incorrect, this is a rather limited view of the nature of germs in the world. For the most part, germs are no different from other creatures that share the earth with us—including insects, which also are generally maligned. Many of us see insects and germs as our mortal enemies, continual nuisances at best, or even predators. But of course, insects are simply looking for food and shelter wherever they can find it. The fact that they tend to like the stores of human food—which includes, for some, human blood—as well as the relative comfort of human abodes, does not make them heinous in any way. *They are doing nothing wrong*. Yet, because humans tend to see insects as simply pests trying to ruin our lives, we refuse to look at them in a good light and ignore their benefits to the ecology.

The same can be said of germs. If they were out to destroy us, they would have succeeded a very long time ago. Yet the human race continues to progress—and even grow—meaning that our relationship with germs is not one of life or death, but rather one of coexistence. Yet until we can appreciate how germs live, what their real code might be as well as how it is used, we have really no chance at reaching any kind of true harmony.

As the Germ Guy, I've tried to bridge the gap between science and society by taking a calm approach to germs and how they relate to us. My main focus has been to dispel the myth of the killer pathogen while being sure to keep people aware that germs can cause infection and that, in a small share of cases, there can be complications and even death. But one message that seems to continually be avoided in the media, as well as in my discussions with people on the street, is the actual reason behind our misunderstanding of what germs mean to us. With this in mind, I've written this book to help smooth out relations between humans and germs. The text is written to provide an understanding of why we seem to continually be at odds with them and how we can look at them in a different light and move forward towards a better coexistence.

While my goal throughout the book is to emphasize balance, there can be no denying that knowledge of—and, unfortunately, attention to—the role of germs in our lives has tended to come from the ways in which they make us sick. For that reason, a large portion of the book is devoted to germs that cause illness. The days of plague, smallpox and cholera still stand as the most memorable moments in our relationship with germs and are the basis not only for the declaration of war but also the development of weapons to fight them. But the battle is unending, marked by the influenza pandemic of 2009 and other worrisome outbreaks such as SARS, *E. coli* O157:H7 and West Nile virus. The impact of infectious diseases isn't limited to our health; our social behaviour and laws have also been dictated by germs of the past and present. Our social mores have changed over the centuries thanks to worldwide outbreaks of sexually transmitted infections such as syphilis, gonorrhea, herpes and HIV.

Yet, despite all the negatives, only a fraction of the human population has really been affected in a damaging way. The majority of us have a relatively calm bond with germs, and as we gain even more insight into their interaction with us, we are learning that our health actually depends on them. Each and every one of us has our own germ populace, known as the *microbiome,* which only recently has been shown to help keep us balanced in an imbalanced world. The microbiome is so critical to health that even the slightest disruption in its population can have drastic consequences, including chronic diseases. Our knowledge of the good germs in our system has led to the development of new means of improving health by manipulating and exploiting the germ code for our medical and corporate benefit. We have germs to thank for everything from antibiotics and insulin to healthy products we use in everyday life such as antioxidants. Right now, this perspective may seem oddly inappropriate, but over the course of this book, the reality that our lives are better *because of* germs will hopefully become evident.

I realized during my research that, in all of the works related to building human relationships with the environment, no writer was more lucid than the great physician Hippocrates. His collections were designed to help people understand their environment, not through preaching or didactic offerings, but through the telling of stories he had encountered over his experiences and relating them to everyday life. In the same way, this book is a collection of stories ranging from crowning achievements in microbiology to more mundane examples of people's individual experiences with germs. I have even included one of my own rather unfortunate experiences with germs.

To keep the focus on the goal of fostering a better relationship with germs, I've tried to avoid academic jargon. I've also forgone the process of including citations and endnotes, but have included enough information to allow you to do further research, should you wish, using your favourite search engine and other reference tools such as Pubmed.

Once you put the book down, the germs will be there, waiting to interact with you for good and for bad. But unlike before you started reading, it is my hope that you will have an understanding and appreciation of the germ code and its application and know how to develop as close to a harmonious relationship with them as possible.

Staphylococcus aureus

1.

CRACKING THE GERM CODE

What's the best way to keep microbiologists happy? Put them in a warm, dark place and feed them decomposing matter. This old joke is, of course, a terrible misrepresentation—most microbiologists I've polled prefer moderate lighting with fresh donuts and coffee—but the other half of the analogy works perfectly. Germs, irrespective of their nature, need only two things to live and prosper: an environment that promotes harmony among their metabolic processes, and a rich, constant supply of nutrients to keep those processes running smoothly.

Given those conditions, microbes can grow incessantly. Take, for example, the bacterium *Escherichia coli,* more commonly known as *E. coli.* With an unlimited supply of nutrients, including decomposing material, fecal matter and other indelicacies, as well as an environment that is both warm and dark—in this case, 37 degrees Celsius and about 75 per cent humidity—a single bacterium could grow and multiply sufficiently to cover the entire surface area of the planet in thirty hours. That's faster than humans can circumnavigate the globe by air (taking layover times into account).

It sounds unbelievable. But the microbiological math that makes it a reality is the true wonder behind these little creatures.

A single *E. coli* bacterium is extremely small—approximately five square microns. A micron is one millionth of a metre, or as it's written scientifically, 1×10^{-6} m. Five square microns, therefore, would be 5×10^{-12} square metres. The surface area of the earth is known to be 510,072,000 square kilometres, which translates roughly into 5×10^{14} square metres. In order for a single bacterium to cover the planet, it would have to somehow multiply some 100 septillion times (that's a one with twenty-six zeros behind it).

Here's where it gets interesting. In these optimal conditions, *E. coli* can reproduce a second generation in twenty minutes, as compared to the nearly twenty years that human beings require for a reproductive generation. Each cell produces two identical bacteria, known as daughter cells, which grow and function in exactly the same way. This process, known as binary fission, is akin to cloning. Within an hour, that one little bacterium of five microns squared has become eight, covering forty microns squared. Following the rule of binary fission, in which the number always doubles (mathematically shown as 2^n), it would take another eighty-six binary fissions to reach that goal of 100 septillion. The amount of time needed would thus be just under thirty hours.

Thankfully, this scenario is merely hypothetical, and germs are not blanketing the entire surface of the earth. Still, they are by far the most prevalent organisms on the planet.

Germs have an accumulated biomass equal to or greater than all the plants in the world. There are more than two billion different kinds of species, and they can be found in every corner of the planet. Closer to home, germs make up about 90 per cent of the human body's cellular composition—most notably in the gastrointestinal tract, but living in other places, from the skin to the lungs to the eyes. The majority of these germs have developed a harmony with this environment to allow mutual growth without the onset of illness or disease. Understanding how these germs have developed this relationship is the basis behind the germ code and the underlying goal of microbiologists.

EACH GERM IS A TINY BUILDING

Making a shelter is common to all higher organisms. Birds have nests, insects their hives and hills, mammals their dens and caves, and human beings our own unique range of constructions. Thousands of years ago, the Hittites lived in the most basic huts of mud and sticks, while the Incas developed majestic cities using merely fieldstones and adobe. Medieval Europe was replete with brick buildings that included majestic castles, while industrialization brought on the age of concrete. Today, urban jungles abound, with giant skyscrapers made from a plethora of different materials, including plastic, glass, synthetic fibres and steel alloys. Modern buildings also showcase other design features, such as controlled temperature and airflow, electricity, water distribution and waste removal technologies.

The main function of any shelter is to maintain a constant and comfortable living environment separate from the outside world. While many basic shelters have only limited success in maintaining this exclusion, modern buildings now allow the possibility of living in an entirely closed space without ever needing to step outside.

These incredible modern achievements in housing have, strangely enough, mimicked the most basic characteristics of germs. At the structural and functional levels, there are too many similarities to ignore.

A building comprises three main components: the outer walls, the inner workings of the people inside, and the individuals who oversee all the operations and make changes when necessary. In the microbial context, the walls are known as the cellular membranes, while the inner workings are collectively known as the cytoplasm. The residents are the various worker molecules needed to sustain life, including those involved in operations, known as proteins; structural engineers, named scientifically as fatty acids or lipids; and the energy and heat producers, called sugars. The most important occupant, much like the monarch in a castle or the prime minister inside Parliament, is the genetic material. Consisting of one or several strands of deoxyribonucleic acid (DNA) or ribonucleic acid (RNA), the genetic material acts as a generating station for the worker proteins and serves as the decision-maker for future reproduction and propagation. While all three components are necessary for life, the importance of genetic material to the building cannot be overstated, as evidenced by an experiment conducted in 2010 by a group of researchers in the United States. They developed synthetic genetic material and put it into a cell whose genetic material had been removed. The new genetic material immediately took up residence in the building and started a new decree, making workers that looked and acted differently and changing the rate of propagation.

Understanding the dynamics of the microbial construct as it pertains to the germ code is a research goal common to all microbiologists. Yet studying these creatures in conditions such as those favouring the aforementioned global *E. coli* blanket only offers a small idea of the microbial realm. To appreciate fully how adept germs are at making themselves comfortable in even the most inhospitable conditions, we must take a look at a small, select group of germs called *extremophiles*—which, as the name implies, have found a home in the most extreme terrestrial environments.

Extremophiles have adapted some pretty clever methods to survive—and thrive—in these environments. And there are microbiologists who, like the great explorers of the past, are driven to venture into these extreme environments in the hope of finding extremophiles. Their goal is to understand the successes of the microbial building architecture to better comprehend how normal microbes work and how humans can better coexist with them. They believe that deciphering these innovative survival strategies can enable improvements in human life in both hospitable and inhospitable conditions. There is also a deeper, underlying aspiration of these researchers in that, as with all explorers, they might find a new species that will be named after them.

STRENGTHENING THE WALLS

In 1967, Thomas D. Brock, one of microbiology's greatest contributors, ventured into the Lower Geyser Basin of Yellowstone National Park, hoping to learn more about microorganisms that are able to survive in incredibly hot conditions. The Lower Geyser Basin is a twelve-square-mile geothermal area in the northwestern United States known for its plethora of hot pools and springs as well as boiling mud pots, which look exactly like they sound. The temperatures can reach upwards of 90 degrees Celsius in some places, and the concept of life would be near impossible. Despite these conditions, there were consistent reports of blue-green algae growing on the surface of certain hot pools.

Brock found that these so-called "algal mats" were comprised not of algae, but rather of a kind of bacteria known as cyanobacteria because of their ability to photosynthesize and thus give off a blue-green colour. When he took them back to the lab, he found that they grew quite well at temperatures up to 75 degrees Celsius. Prior to this discovery, the highest temperature for any living organism was thought to be 60 degrees Celsius.

This discovery alone was enough to complete his project (and presumably fulfill his funding objectives), but he wasn't finished. He believed that if he could find bacteria growing happily at 75 degrees Celsius, then perhaps he could discover something living at even higher temperatures closer to the basin.

The geysers in the Lower Basin are as well known for the consistency of their water and steam eruptions as they are admired for their striking pink and white streaks. Traditionally, the colour was assumed to be the result of high mineral content from the spewing geysers; however, upon closer investigation, Brock found that the pink and white optics resulted from the presence of an unusual bacterium. These filamentous bacteria could only be seen under the microscope, but their abundance caused the ground to exhibit a particular hue.

It took a few years to learn how to cultivate the bacteria in the lab—they seemed to only enjoy the unique conditions of the Geyser Basin—but with patience and an infinite number of culture media mixes, Brock managed to grow the bacterium in the lab. After giving it much thought, he decided not to name it the ignoble choice of *Thomasus brockii;* instead, he chose *Thermus aquaticus*.

With the successful isolation and propagation of *T. aquaticus*, Brock and others were free to start learning the secrets behind this bacterium's ability to survive in such a harsh environment. He learned that the secret wasn't all that different from certain aspects of modern architecture, albeit designed millions of years earlier.

Heat is a natural process caused by the continual movement and bumping of molecules. It is self-sustaining in that, as it increases, it causes molecules to move more quickly and collide. Consequently, increased heat puts two forms of stress on complex molecules, known as shear and stretch, and when the temperature rises to a certain level, complex structures begin to break at the molecular level. As the temperature rises to extreme levels, solids turn to liquids and, without cessation, those liquids inevitably turn into gases.

Under normal conditions, the walls of a building are sound structures, yet in the presence of massive shear and stretch, such as those found in a fire, the walls are put under significant stress and collapse. At the microscopic level, the same kind of shearing and stretching happens to a membrane at high temperatures; without some kind of reinforcement, the membranes will simply rip, tear and dissolve, killing the cell.

T. aquaticus overcomes this problem through a rather interesting membrane-design alteration. Instead of long, thin lines, the membranes are cross-linked and pouched, providing more complex interactions between the fatty acids and proteins. The walls are therefore thickened to protect against shearing, while still maintaining flexibility to withstand stretching. Perhaps more impressive is the fact that the membrane changes its structure dynamically depending on the temperature. The bacterium can live as comfortably at 50 degrees Celsius as at 90.

The complex membrane structure of *T. aquaticus* has since been adapted by architects to help improve wall stability and tensile strength. Today, most buildings are designed with shear walls, which incorporate a cross-linking matrix of rebar to resist movement caused by heat and other stressors, such as earthquakes. Moreover, the use of pre-stressed concrete, which involves adding flexible tendons that are stretched and compressed to test and confirm the elasticity of the structure, will resist stretch phenomenon. While human-made structures may never see the same conditions as their microbial counterpart, its contributions will help create stronger buildings in the future.

KEEP THE INHABITANTS WARM AND HAPPY

As almost anyone who has lived in a temperate zone understands, warmth is an issue when it comes to being able to function. When the temperature drops significantly—especially below the freezing point—everything literally slows down or stops. The answer in modern-day society has simply been to "turn on the heat" or, for longer stretches of cold, to increase the insulation within a building to keep the cold out. Unfortunately for microbes, there are no heating elements to rely on. As a result, they have developed some fascinating means of ensuring that the population inside the cell remains comfortable.

The quest to find and understand germs in the Arctic—known as *psychrophiles,* or cold-loving microbes—began back in 1868 when a three-man group led by the Finnish arctic explorer Adolf Erik Nordenskiöld ventured out from their home base in Karlskrona, Sweden, on a four-month journey aboard the steamship *Sofia* in search of the North Pole. At one of the stops—Spitsbergen, Norway—one of the team members, Dr. C. Nystrom, collected soil and sea samples to determine whether he could detect microbes. What he found was that, despite the near- or below-freezing temperatures, he could find fermenting and putrefying bacteria. The levels were sparse at best, but the news sent waves throughout the microbial world and initiated the hunt for cold-loving bacteria and to understand how they survive.

Over the last century and a half, thousands of researchers have since braved the subzero temperatures of both the Arctic and Antarctic to collect microbial samples. They have resorted to some rather ingenious techniques, ranging from the use of modified floor mats to the attachment of petri plates to the nose of an airplane. What they have found reveals a marvel of microbial adaptation.

The first alteration is similar to inner-wall insulation, albeit on the outside of the cell. This protective layer is made not of fibreglass or polyurethane but of exopolysaccharides (EPS), which are essentially sugars that function to release heat to keep the internal temperature high enough to allow normal function. Should the temperature drop even further and the EPS be unable to keep the cell warm enough, these bacteria have developed two other strategies to keep the cold at bay. The first is to continually produce proteins that prevent ice from forming. These aptly named "anti-freeze proteins" are specifically designed to identify any nanocrystals of ice that may form within the cell, attach to them and reliquefy them.

The second and more ingenious strategy achieves what was once considered to be impossible. These oddly cold-loving cells somehow manage to change the design of a group of co-workers inside the cell, known as *enzymes,* so that they work better at lower temperatures. Enzymes are one of the most important operational proteins in the cell and are mandatory to sustain life. They are the project managers of the cell, working to bring molecules together to create a product, and are easily identified by the use of the last three letters *-ase;* an enzyme that breaks down proteins is known as a *protease,* while an enzyme that produces polymers of molecules is known as a *polymerase.*

Each enzyme has a specific molecular structure, and the

dogmatic belief has been that even the slightest change in that structure could severely alter or even quash the project-management function. Yet, in certain cases, this rule has been thrown out the window.

Take, for example, amylase, which is common in almost all living species and is abundant in human saliva. The enzyme, as its name implies, breaks down starches (known scientifically as amyls) to simple sugars that are necessary for proper metabolism. Normally, amylase is only effective above zero degrees Celsius, but certain bacteria in the Antarctic have enzymes that are functional even at minus-20 degrees. When researchers tried to identify how the enzyme worked so effectively, they found that the structure had been modified by an amazing and unheard-of one-third. The observation flew in the face of dogma, and even today, it confounds researchers as to how this type of evolution is even possible.

The wonders of the psychrophiles have contributed to a tremendous leap forward in the understanding of terrestrial ecology, and the identification of cold-adapted enzymes such as amylase has also led to several advancements for human use. Similar enzymes are now added to laundry detergent to increase the effectiveness of cold-water washing. Other enzymes are being used to reduce the need for higher temperatures in the fermentation process for cheeses and bakery items. But perhaps more important, these bacteria are the basis for an entirely new generation of bioremediation, where otherwise potentially toxic chemicals can be replaced with less harmful microbes, and the effect of temperature on a project is of limited concern.

THE KING IS DEAD; LONG LIVE THE KING

In collective communities, whether composed of humans or bees, one fundamental rule is that the leader must be protected to maintain order and prevent anarchy. In the beehive, the queen is surrounded by more than a dozen bees that spend their lives feeding, grooming and protecting her from outside elements and would-be conquerors. She in turn safeguards the hive, so that it functions normally, and lays her eggs to keep the population thriving. In the human context, civilization is based on the hierarchy of the political structure, which is ruled by kings, queens, presidents and prime ministers. To maintain order, the safety and welfare of these people must be protected, and countries employ hundreds if not thousands of individuals who will risk their lives to keep anything untoward from happening. In turn, the leaders do their best to ensure that the land prospers, the economy grows and the people feel free to live their lives in comfort and have children who will enjoy a similarly harmonious life.

For a germ, order is dictated by the genetic material, which contains the code for survival through the proper production of workers and through replication to ensure propagation of the species. Depending on the type of microbe, the genetic material is normally protected either through a covering layer of worker proteins, much like the queen bee, or through an intricate protective superstructure that hides the genetic material from outside harm. These strategies are mainly effective, but there is one enemy that can overcome all these protective measures and harm the genetic material permanently: radiation.

We've all heard that radiation is a threat to health and that exposure could lead to drastic consequences such as burns, cancer and death. Radiation affects any living creature by damaging or destroying the genetic material. Upon exposure to radiation, DNA undergoes a rather nasty process where some of its building blocks, known as *nucleotides,* fuse together. This fusion puts undue stress on the genetic material and, much like the shear and stress on a membrane on buildings, leads to breakage. These fusions and breaks prevent the genetic material from coding the necessary proteins for life, and the cell inevitably dies.

To counter this, microbes have developed molecular mechanisms to repair and heal the damage, not unlike a doctor coming to the rescue. The most understood of these rescue measures is what is known as the SOS (save our ship) response. An enzyme known as a *recombinase* first finds the wounded genetic material and cradles it in a molecular pocket. Next, a second enzyme, a polymerase, comes to remove the damaged section and encode a new one. Once the new strand is made, the recombinase sets the DNA back in place and ensures that the code of the genetic material has not changed.

This process was discovered in 1975 by Miroslav Radman, who has since devoted his career to finding the means of immortality. His theory of life extension through the preservation of genetic material suggests that the keys to longevity exist within the understanding of what he calls the "biology of robustness"—the investigation and identification of pathways that contribute to aging and means of preventing them from happening so quickly. Thanks to a little-known but extremely "robust" bacterium, Radman may one day see his dream come true.

Deinococcus radiodurans is what is known as a *polyextremophile.* Its name alone is dramatic—it literally means "strange berry that withstands radiation," although the bacterium can do so much more. It can survive extreme heat and cold, as well as the pressures of a vacuum, the corrosive effects of acid, and high levels of the normally deadly mercury. It has been named the world's toughest bacterium by the *Guinness Book of World Records* and is probably one of only a few life forms that could survive an apocalyptic event. What makes this bacterium so interesting to Dr. Radman is that it possesses a trait unseen in any other species: it can resurrect its king from the death throes of lethal radiation.

Radman used a classical microbiological technique in which a microbe is subjected to a stressful situation and the reaction is then observed. He administered a dosage commonly used in food irradiation—7,000 grays, which is equivalent to about ten times the lethal human dose—and then left the bacteria to die. At the molecular level, the genetic material was shattered into several pieces that normally would be so dysfunctional that the cells would, in essence, be clinically dead.

Shortly after that exposure, though, something incredible happened. First, the cell immediately called in a number of proteins that huddled around the small pieces of DNA and protected them from any further degradation. Then, another set of proteins matched pieces together as if solving a jigsaw puzzle. As soon as an appropriate match was found, the polymerases came in to create new strands. The process continued until all of the genetic material was restored and the microbe could once again function normally.

The incredible regenerative process of *D. radiodurans* has been attributed to the evolution of the bacterium to adapt to very dry conditions, such as those found in deserts and in permafrost. In both these environments, solar radiation is extremely high, and without a highly structured repair mechanism, any organism that finds itself on the surface would meet its demise. This history has caught the attention of many astrobiologists who continue to search for life in space—more specifically, on Mars. While some may feel that *D. radiodurans* could have originated on Mars, most prefer to believe that it is still terrestrial but a model for the "training" of species to resist the Martian-like conditions of low temperature, low humidity and high radiation. If other types of soil bacteria can be trained to exist and thrive in such conditions, the potential for recolonization of the soil is apparent, as is the potential for the initiation of an agricultural society based on Martian soil rather than our own earth.

THE REVELATION

The world of extremophiles continues to fascinate microbiologists, and future discoveries will no doubt bring us closer to understanding how these tiny "buildings" survive and how humans can utilize these innovations to improve life here on earth and perhaps beyond. The true contribution of these studies, however, is the deciphering of the germ code and revelation of the basis of all things germy:

In order to survive and thrive, germs evolve and adapt.

In the human context, this realization has led to a deeper understanding of how germs colonize and flourish in the warm, moist and nutritious environment the body provides. This knowledge also sheds light on the ways the approximately 1,450 species—including *Campylobacter*, influenza, norovirus, cholera, tuberculosis, SARS, dysentery, yeast and certain strains of *E. coli*—have learned to cause infection and led to occasional deleterious effects. To maintain human health, there has been little option but to fight back, though with such a significant disadvantage in terms of the ability to evolve on the biological front, an assortment of antimicrobial weapons has been required. A plethora of antibiotics, disinfectants, antiseptics and other treatments has been designed and used to keep the bugs at bay. But, as history has shown, germs inevitably find a way to evade these weapons and continue to find a happy, human home.

2.

A WAR WE CANNOT WIN

Wars are an interminable part of human existence. We have waged them throughout our history and show no signs of giving up. In fact, we have broadened the field over the last century so that opponents are no longer defined only as nations or populaces, but also as abstract ideas or conditions. These wars without leaders or targets include three declared by American presidents: Johnson's "war on poverty" in 1964, Nixon's "war on drugs" in 1971, and George W. Bush's post-9/11 "war on terror."

Taking on an unseen enemy has great value in gaining public attention and engagement. But the downside is that victory is, shall we say, unlikely. Poverty and drugs are more prevalent in society today than when they were first targeted. As for terrorism, George W. made his "mission accomplished" announcement in 2003, but the enemy wasn't listening.

The "war on germs" began with the emergence of our species, although it wasn't until a few decades ago that victory was supposedly announced. In the late 1960s, the U.S.

surgeon general William H. Stewart was purported to have said: "It is time to close the book on infectious diseases, and declare the war against pestilence won." This never-documented but widely quoted proclamation sent shock-waves through the global community, particularly to those who were actually fighting germs. To them, there was no tangible evidence to suggest that the ongoing struggle with germs was resolved. There might have been victories along the way, but no one in their right mind would ever have asserted victory. When asked about it thirty five years later, Stewart said he couldn't remember whether he'd had this "mission accomplished" moment or not.

Much like any war, the one on germs has its proponents who want to wipe out all germs (an ideology more than anything else, as it is impossible to rid the world of its microbial population); its detractors, who believe that coexistence is better than battle; and those who believe that germs are not worthy of a war simply because germs are everywhere and cannot be controlled. While no one camp will ever gain the upper hand, there can be a harmonized vision that incorporates all three objectives. A selective approach allows the ability to determine which agents are the targets for eradication, which ones deserve appreciation and thanks, and which ones are simply not worth the consideration. Once the germs that are worthy of a battle to the death are identified, a strategy for engagement can be developed. As with other great leaders who have gained victory over a difficult opponent, we can turn to the past to guide us, so that winning efforts are maximized and losses are minimized. In this context, there is no better guide than a man who spent his entire life fighting not one, but a plethora of enemies and never once failed. This man was the ancient Chinese general, Sun Tzu.

THE ART OF WAR

Sun, a military general in China during the latter part of the sixth century BCE, was lauded for his conquests during an unsettled time in that nation's history. He was primarily a strategist who focused on gaining the greatest advantage over any opponent, and then carrying out his plans with astounding success. He passed on his knowledge in *The Art of War*, a concise book of fewer than 100 pages. It was valued throughout several Chinese dynasties and is still consulted today by anyone—generals and businesspeople alike—seeking to defeat an opponent tactically rather than through simple brute force.

The text comprises thirteen chapters, ranging from laying plans before battle to identifying the best tactical positions and the best use of terrain. The words are simple to understand and precise in their meaning. Still, academics have distilled Sun's words further in a modern point-form strategy. Five of those points work very well as a means of analyzing our past, present and future struggle with germs:

- *Know Your Enemy*
- *Know Your Weaknesses*
- *Know Your Weapons*
- *Attack in a Concerted Way*
- *Beware of Lengthy Campaigns*

Know Your Enemy

Humans have a habit of attributing the unknown to the supernatural, and for thousands of years infectious disease was thought to be caused by the actions of a deity or demon. In the fourth century BCE, the ancient Greek physician Hippocrates tried to dispel this belief by offering a more scientific explanation for illness. In the treatise *Airs, Waters and Places*, he claimed that sickness was associated with climate, soil, water and lifestyle, that certain diseases were commonplace, or endemic, to a region and that, on occasion, there were outbreaks, which he called "epidemics." Hippocrates also listed several maladies, including dysentery, fever, meningitis, rashes and pneumonia, all of which have since been associated with an infectious agent.

Despite his insightful writings, Hippocrates faced an all-too-common problem in that he didn't mention an actual agent of disease. Without the ability to see, touch or manipulate an offending party, the declarations, no matter how sound, were still regarded solely as theories and therefore not worth developing into policy. As such, the world continued to believe that disease was primarily associated with an unknown entity. Two thousand years would pass before a simple Dutch draper with a skill for grinding lenses would provide the evidence necessary to prove Hippocrates correct.

In 1654, twenty-two-year-old Antoni van Leeuwenhoek started his business in Delft, a small community in the Netherlands. Draping, however, was not enough to keep him engaged, and he sought diversions to occupy his spare time.

One such hobby was grinding glass into a magnifying hemisphere. Leeuwenhoek found that his precise and exacting approach to draping translated well into grinding lenses, and soon he had a decent collection that could magnify objects between fifty and five hundred times. With this new tool in hand, he did what any burgeoning researcher would do: he started looking at really tiny objects. He spent close to a decade looking at solid matter ranging from human hair to the white, chalky cliffs of England. In 1676, he changed his focus to rainwater, if only to see what might be eroded during a storm. What he found surprised him and changed the war on germs forever.

Underneath the lens was a collection of single-celled organisms. He called them "animalcules"—literally, "little animals"—and proceeded to send his findings to the British Royal Society, which published them in its journal, *Philosophical Transactions*. His work sparked significant interest across Europe and opened the door to the microscopic investigation of life, better known today as microbiology. While curiosity raged, another two hundred years would pass before these tiny creatures would be identified as the cause of infectious disease.

In the mid-1800s, Louis Pasteur was a professor of chemistry in Lille, France, and was working to identify how a certain salt bent light. In 1856, in the midst of this somewhat monotonous job, Pasteur was approached by a local distiller requesting his help in solving a rather important business problem. The distillery produced alcohol from beets, but its product was consistently going off and turning sour. Pasteur agreed to help and used a microscope to identify the animalcules in both the normal and ruined batches. What he found was that, in addition to the usual round microbes associated with fermentation—what we characterize today as brewer's yeast—the bad batch had a collection of other tiny living beings present in relatively high numbers.

Pasteur went on to find these beings, which he now called "germs," in almost every environment he tested, including the air, water, other surfaces and, most important to him, human skin infections. He called this microbial cause of illness the Germ Theory and shared his findings with the scientific community. Soon, other researchers were looking into everything from the pus in wounds to the brains of cadavers, all in the hope of finding tiny germs that could cause disease.

One such researcher was Robert Koch, who spent the better part of the 1870s and 1880s focused on a specific disease: anthrax. He spent more than a decade figuring out how to develop lab cultures of the responsible microbe, *Bacillus anthracis*, and then tried using these cultures to infect mice. His success allowed him to further investigate the association between germs and illness. His results were so concrete that, instead of theories, he developed postulates for microbial disease. These requirements, which included not only the identification of the germ in a diseased host, but also—and more important—the transfer of the disease to another host through infection with a separate culture of the microorganism, became the basis for infectious-disease research and the threshold that germs need to achieve in order to be called what we know today as *pathogens.*

More than 1,450 different kinds of pathogen have been

identified, and their numbers continue to grow. Almost all of them have been classified into one of five general categories of germs, and then given a particular species name depending on certain specific traits. The categories refer to the overall general structure of the microbial buildings, while the species provides a link to a specific characteristic or trait associated with its structure or its inner workings.

Bacteria are the best-known category of germs and comprise the greatest number of species. There are close to two billion different species, most of which are completely harmless to humans. They are only a few millionths of a metre in length and can survive on their own in the environment. They come in a few general varieties, but the majority are either spherical (known scientifically as *coccus*) or rod-shaped, which at one time was denoted as *bacillus.* There are a number of bacterial pathogens; however, a select few grab the most headlines or warrant mention in this book. They are summarized in the table below.

Name	Pronunciation	Meaning	Disease(s) caused
Staphylococcus aureus	staff-ill-oh-kok-us aw-ree-us	Spherical bunches of golden grapes	Skin and respiratory infections, pneumonia, flesh-eating disease
Escherichia coli	esh-uh-reek-ee-ah koh-lye	Theodor Escherisch's colon	Gastrointestinal illness, diarrhea, urinary tract infections
Neisseria gonorrhoeae	nye-seer-ee-ah gawn-oh-ree-aye	Albert Neisser's flowing seed	Gonorrhea, pelvic inflammatory disease, sterility, blindness

Name	Pronunciation	Meaning	Disease(s) caused
Mycobacterium tuberculosis	my-koh-bak-teer-ee-um too-ber-kew-loh-siss	Slimy bacterium that causes pimples	Tuberculosis
Salmonella typhi	sall-moh-nell-ah tye-fee	Dr. Daniel E. Salmon's stupor caused by fever	Typhoid fever
Campylobacter jejuni	kam-peh-lo-bak-ter je-june-ee	Curved rod that affects the jejuni of the intestines	Gastrointestinal disease
Bacillus anthracis	bah-sill-us an-thray-siss	Stick that causes boils or carbuncles	Skin and respiratory anthrax
Streptococcus pyogenes	strep-toe-kok-us pie-ah-gen-ease	Twisted spheres causing pus	Scarlet fever, sore throat, pneumonia, meningitis
Clostridium difficile	kloss-trid-ee-um dif-fi-seal	Spindle that is difficult	Chronic gastrointestinal illness, diarrhea, pseudomembranous colitis
Yersinia pestis	yer-sin-ee-ah pess-tiss	Dr. Alexandre-Emile-John Yersin's pestilence	Plague
Aeromonas hydrophila	air-oh-moan-az hi-dro-fill-ah	Gas-producing units loving water	Diarrhea, flesh-eating disease
Vibrio cholerae	vihb-rio ko-ler-ay	Curvy organisms that cause nausea, vomiting and diarrhea	Cholera

Viruses are significantly smaller than bacteria, with the majority ranging in size between 30 and 300 billionths of a metre (that's 10^{-9} metres!). Only about five thousand types have been discovered to date, but they make up for their relative lack of diversity with an incredible ability to do harm. Unlike bacteria, which can live independently and theoretically cover the planet, viruses are strictly intracellular pathogens. They need to be inside another structure, such as a human, plant or bacterial cell, in order to survive. Once there, however, these little machines can reproduce at a lightning pace, producing several hundred daughter viruses in a matter of a few hours and upwards of tens of billions in about a day. This kind of rapid generation makes viruses a great problem in terms of health because a rapid response is needed to prevent the most harmful effects. Unfortunately, in many cases, treatments are just not available and little can be done.

The viruses that keep us awake in the headlines are listed in the table below.

Name	Meaning	Disease(s) caused
Rhinovirus	Virus of the nose	Common cold
Influenzavirus	Influential virus	Flu
Coronavirus	Crowned virus (from its shape under the electron microscope)	Common cold, SARS
Norovirus	Norwalk virus	Diarrhea, vomiting
Human Immunodeficiency virus	Virus that impedes the immune system	Immune dysfunction leading to AIDS
Ebola virus	Virus from the Ebola River valley in the Congo	Hemorrhagic fever
Rabies virus	Virus that causes rage	Rabies
Dengue virus	Prudery virus	Dengue fever, including stiffness of joints
Hepatitis virus	Virus that causes disease of the liver	Gastrointestinal disease, liver disease, immune dysfunction

Fungi—yeast and moulds—are a special group of germs. As unicellular organisms, they are too small to be seen by the human eye, but many have the ability to create visible superstructures. The classic example of a superstructure fungus is a mushroom. Fungi have an interesting place in the war because, for the most part, they are allies rather than enemies. Even before the visual identification of germs, fungal species were used in the fermentation of foods and drinks to develop alcohol or vinegar; the process was determined to be a matter of an unknown entity that somehow used sugar and milk products to make beers, wines and yogurts. The pathogenic potential of fungi was once limited to vaginal and skin infections, but with the advent of AIDS, several fungi found a home in infected individuals who had a compromised immune system. Names such as *Cryptococcus* and *Histoplasma* became commonplace, and many AIDS patients succumbed to these uncommon infections. Still, out of the estimated five million species that exist, only a handful has led to human disease.

Protozoa—"first animals"—were the principal constituents of Leeuwenhoek's microscopic investigations. These microbes find a home in almost any wet environment and can be found regularly in water sources and even the blood of certain insects, such as mosquitoes. There are only a few dozen protozoa that actually cause human infection, but the effects can be disastrous. The main protozoan pathogens are featured in the table below.

Name	Pronunciation	Disease caused
Plasmodium vivax	plaz-moh-dee-um vye-vax	Malaria
Entamoeba histolytica	en-ta-mee-bah his-toh-lit-ik-ah	Dysentery
Naegleria fowleri	nae-glee-ree-ah fow-ler-eye	Panamebic meningoencephalitis
Giardia lamblia	gee-ar-dee-a lam-lee-a	Gastrointestinal disease
Cryptospordium parvum	crip-to-spor-ih-dee-um par-vuhm	Gastrointestinal disease, chronic watery diarrhea
Trypanosoma cruzi	tri-pan-oh-soh-mah kru-see	Chagas disease

Name	Pronunciation	Disease caused
Acanthamoeba castellanii	ay-kanth-a-mee-ba kas-tell-a-nee	Eye infection that can lead to blindness
Leishmania major	leesh-may-nee-ah	Chronic open wounds that do not heal
Toxoplasma gondii	tox-oh-plaz-mah gone-dee-eye	Mental illness, encephalitis, abortions

While not entirely microscopic, *helminths* are parasitic worms that find a happy home inside the body. There are only three types of worms that can cause disease in humans, but their impact is rather astonishing. *Trematodes,* also called flukes, are leaf-shaped worms that have a similar appearance to a leech. Once in the body, they can reside in the gut or migrate into the blood or to a preferred organ such as the liver, pancreas or bladder, where they attach to the organ and feed. The most common flukes are the *Schistosoma* species, which causes overall weakness as well as diarrhea and abdominal pain. *Cestodes* are tapeworms, and one need only perform an Internet image search to see what happens when there is an infection. The tapeworm resides in the gut and begins to grow as it feeds on digested food passing through the intestines. If left undiagnosed, the tapeworm can grow to an incredible length of well over ten metres. *Nematodes,* better known as roundworms, make up the final group. These tiny worms also reside in the gut and, akin to other bacteria, multiply quite rapidly. Infection can lead to fatigue, diarrhea, nausea, weight loss and reduction of immune activity.

There are also pathogens with no building structure—individual workers that have somehow ventured off on their own. The prion is a prime example of this kind of rogue worker. This protein comes in two forms: normal and variant. The normal form finds a home in the cell and helps to maintain structure and integrity. The variant form wants nothing to do with the cell and resides on the outside, known as the *extracellular matrix*. The prion enters the body, usually through contaminated food such as meats from infected—more commonly known as mad—cows, sheep and deer. Rather than being digested like all other nutrients, the insurgent travels from the gut to the blood to the brain. There, it finds a comfortable place in the extracellular matrix and starts changing other prions from helpers to attackers. At the molecular level, this process involves a chemical alteration of the normal prions such that their shape and function mimic that of their new leader. As the process—which can take years—continues, the cells that depended on the normal prion workers begin to die and fade away. The result is an open space in the brain called a plaque, which leaves the person with a gap in their normal cerebral function. As the plaques grow and spread, they begin to shut down the body's ability to think and take care of itself. The infection is terminal, and there is no cure.

Know Your Weaknesses

With germs identified and classified, the next stage in winning a war against them is to identify the ways in which our society allows germs to infect, cause disease and kill. Perhaps the best place to begin is with the dawn of humanity, and how our ancient human ancestors lived with—and died with—germs.

Paleontologists are unique individuals who strive to provide clues to the life and times of ancient humankind. Within this group are a devoted number of people known as paleopathologists, who focus on how germs affected our ancestors and the rather unpleasant results of a world without formal medicine. These scientists have an incredible ability to look at bones that are hundreds, if not thousands of years old to find specific markings that identify an infection and, depending on the extent of the injury, the living hell that was endured as a result. From their analysis of remains, several germs have been clearly identified based on distinct markings and deformations of bone.

Perhaps more fascinating—or depressing, depending on your perspective—their work has shown that the germs of the ancients are the same ones causing problems today. The leading killers of the ancients included tuberculosis, anthrax, syphilis, leprosy, sleeping sickness and tapeworms. These germs were distributed globally, and evidence of their scourge can be found on almost every continent. The dates of infections seem to parallel major strides in the progress of human civilization.

In prehistoric Egypt, the development of cities led to great advancements in architecture, socioeconomics, politics and urbanization. But it also led to an increase in sedentism, in which nomadic behaviour gives way to a people settling down in one place. While the Egyptians were no couch potatoes, they did happen to enjoy the sedentary life, and the population soared. As the number of residents grew, so did the level of infectious diseases.

More people require more food, and agriculture became big business. However, the increase in plants being cultivated led to an increase in disease-carrying insects. Livestock numbers also grew, and animals shared their illnesses with their farmers through close contact. There were also higher levels of fecal-based gastrointestinal illness through improperly prepared food and contamination of food by human and animal waste. A plethora of germs have since been identified as having caused infections through agriculture, and many of them continue to haunt us today—from lethal *E. coli* outbreaks to bird and swine flus to mosquito-borne infections such as malaria.

At about the same time that the Egyptians were building the pyramids, the Orient was learning the art of forging metals and entering a mini-industrial age. The levels of infectious disease also apparently increased during this time. But unlike the civilization issues in northeast Africa, these people were fighting off what we consider today to be normally harmless fungal pathogens. The bones speak of a people who had a reduced level of immunity and a higher incidence of disease. Somehow, the actual interaction with metals was leading to a decreased immune system, leaving them more likely to become infected.

These industrial pioneers had no idea that they were suffering from metal poisoning and the effects of pollution. Indeed, this health problem wasn't considered a medical condition until the Industrial Age of the latter part of the last millennium. What became clear, then, is that the widespread forging of metals, as well as the pollution that results from metalworking, interfere with the immune system, altering its ability to function normally. Unfortunately for those in the Orient, a lack of knowledge led to many unnecessary infections and death. But their losses were nothing compared to another society that endured an eternity of death by germs as a result of the lack of one very important human need.

About 500 AD, the Roman Empire was at its height. The Romans had developed the concept of a fully functioning civilization and kept it stable through a combination of architectural advancements and territorial expansion, usually through the invasion and subsequent takeover of adjacent lands. Their greatest achievement, however, was the building of the aqueducts, which brought safe water from faraway freshwater springs to the city. At the height of the Empire, the city of Rome was serviced by nearly four hundred kilometres of aqueducts, the longest of which, the Aqua Marcia, was over ninety kilometres long.

The aqueducts were by far the most ingenious and effective means of maintaining a civilization. They were also the most frequent target of their enemies and the reason behind the fall of the empire. In 537 AD, the Germanic Goths, after almost two centuries of trying to defeat the Romans, decided upon a new tactic to overthrow their Roman enemy. Instead of focusing on the city of Rome, they attacked the aqueducts, destroying them and leaving the city without a source of clean water. The effect was devastating; the lack of drinking water forced many to leave the city and fall prey to the ruthlessness of the Goths.

The loss of the aqueducts sent Rome into the deadly era of the Dirty Millennium. During this time, Rome was known not for its historic empire but for its inability to solve its germy problems. All who dared venture there could expect to suffer some kind of infection. There were various attempts to improve matters, but nothing seemed to work. But in 1453, Pope Nicholas V had a no doubt divinely inspired idea to rebuild one of the aqueducts and bring fresh water back to the city.

Today, clean water may seem like a given for health, but the Romans needed nearly a thousand years to learn that human existence is entirely dependent on the supply of fresh, running water for drinking and bathing. But clean water is nothing if living in filth is the norm. The realization of this fact came in the fourteenth century, when the Great Plague ravaged Europe and killed nearly a quarter of the world's population.

For millennia, the plague had been an innocuous disease that was local to remote areas of the Orient and had little to no impact on civilization. But when the disease travelled along the silk road from China and Europe and happened to befall a group of Tartar warriors during an attempted siege of the trading city of Caffa (known today as Feodosiya, Ukraine), the need for a clean environment was revealed to be paramount to prevent a horrific wave of death.

The plague is caused by the aforementioned *Yersinia pestis,* and it spreads through the bite of an infected flea. The ailment begins with the swelling of the lymph nodes, particularly in the armpits and groin. These nodes can grow to about the size of an egg, leaving the infected person with reduced mobility as well as a high fever. Without treatment—and until the mid-twentieth century, there was none—the bacteria move to the extremities, leading to gangrene and death of the fingers, toes and nose. In a matter of days, death ensues, but not until the body has all but been consumed from the inside out.

When the Tartars experienced the outbreak of the plague, they did something rather unique with their dead comrades, whose stench and appearance made for low camaraderie. Using catapults, they flung their dead into the city of Caffa in the hope that the infection might spread and kill the inhabitants, leaving an easy battle ahead. The strategy worked to some extent, but it did have an unexpected consequence: the residents of Caffa who didn't die became so disgusted with the continuing rain of the dead that they figured life would be better away from the city. They left and travelled to other areas of Europe and the Middle East, taking the disease-ridden fleas with them. Wherever the people landed, the plague would instantly appear.

The Black Death was at its peak for only two years, but the disease appeared sporadically for another three centuries until living conditions improved and humans were finally separated from the fleas. The lesson learned from this momentous period in humanity's war on germs should have been heeded half a millennium later, when the Great War started. If it had, millions of lives might have been saved.

Trench warfare subjected troops to poor diet, questionable water supplies, wet and humid living conditions, overcrowding, absence of proper waste disposal, and a lack of proper wound care. Battle wounds predominated, but close behind in number were the infections, including the infamous gangrene of trench foot, respiratory illnesses, undiagnosed fevers and meningitis. Upon presenting himself to the commanding officer with an illness, a soldier had to travel several miles in treacherous conditions to seek medical help; many were far too late for proper treatment. For them, the only answers lay in amputations—usually without anesthesia—isolation to prevent infection, and ultimately, death. When the war ended, about half of the deaths on either side were attributed to germs.

The Great War was an unfortunate event, although its mark on us sparked a change in the way we live as a populace. We now know that focusing on safe food, clean water, proper sanitation and decent shelter are necessary to maintain a balance with germs. Many countries have achieved these milestones of development, but sadly, these tenets are not universally applied. Today, some 15 million people still die needlessly every year from infections caused by poor living conditions.

Know Your Weapons

Almost a hundred years before Koch's postulates were shared with the world, the identification of weapons against germs was underway, although the intent had more to do with keeping people alive than finding and fighting germs. But in a world where death caused by an unknown enemy was the norm, anyone who believed in a better future would be willing to try anything to gain the upper hand, even if it meant risking the life of a child against one of the most deadly diseases.

Vaccination

The lethality of smallpox is legendary; up to 30 per cent of those who acquired it died, and at its height, the virus killed some 400,000 individuals per year. The infection starts off slowly, causing only a fever and general uneasiness at its onset. But as the infection spreads, the symptoms turn dramatically worse. The skin develops a rash that looks like a burn, and soon the pimples begin to appear. As the virus load grows, the pimples turn into pus-filled growths that rip open, allowing the virus to spread in the environment. At this point, the virus turns inward, heading to the internal organs, causing similar growths and tears on the liver, heart, throat, lungs and intestines. The effect on the body can be just too much for some. For the 70 per cent who survive, life is a struggle as the body becomes deformed, many end up blind, and the individual is left with a daily reminder of the infection with scars from the pimples, called pockmarks.

In 1796, smallpox came to England and left about 35,000 people dead in its wake. The virus was indiscriminate in its socioeconomic scope, killing the rich as well as the poor. Yet amidst the calamity, one segment of the British population seemed to be immune to the deadly throes of the virus. Reports appeared from the Isles that dairy milkmaids were evading infection; the reasoning being that they had already suffered from a bovine version of smallpox, appropriately named cowpox. The news reached the ears of a Berkeley researcher named Edward Jenner, a classically trained biologist turned physician with a penchant for travel in an effort to understand the nesting behaviours of the cuckoo bird and the hibernation of hedgehogs. His interest was piqued, but considering he had settled down in Berkeley with a wife and two children, he realized that his adventurous days were behind him. Then, one day, as if through divine intervention, a young milkmaid named Sarah Nelmes approached him and asked him for a way to help stop the incessant itching and scratching of the rash and lesions on her hands and arms. Jenner recognized the symptoms right away; this was cowpox.

The researcher in Jenner was reawakened, and he knew exactly how to test the hypothesis.

The experiment that followed would easily fulfill the "do not attempt without proper ethical review board approval" requirements, and in today's world it could end up costing one's funding. Jenner scraped some of the cowpox lesion material from Sarah's arms and then transferred it to the arm of a young lad helping the Jenner family with their chores, the eight-year-old James Phipps. Ethically, this portion of the experiment wasn't all that problematic; cowpox wasn't lethal. What happened next was the real breach of ethics. Nine days after the cowpox inoculation, Jenner took lesion material from an actual human smallpox victim and scraped it into James's arm, putting the boy's life at risk.

One can only imagine the shock and fear that James's parents must have felt as they literally risked the life of their son to support what many might claim to be an insane proposition. Thankfully, though, nothing happened. There were no lesions, no fevers, no malaise; not even a rash.

With the potential to save the world from smallpox in hand, Jenner was granted permission by his wife to venture to London to share his findings. While there were detractors—in research, there always are, no matter how sound the evidence—he gained the interest and approval of several physicians who were desperate to try anything to prevent smallpox from taking another life. Over the next two years, the practice spread across the British Isles, and by 1800, it was widely used in continental Europe. That same year, the vice-president of the United States, Thomas Jefferson, was sent the vaccine and had his doctor administer it to him. When he became president later that year, he set up the National Vaccine Institute and began the first widespread vaccination program.

Today, there are vaccines for close to two dozen groups of pathogens, including the measles and mumps viruses, the hepatitis A and B viruses, the bacterium that causes whooping cough, typhoid fever and poliovirus. Vaccination records are a regular part of any health record, and many countries now require a certain level of vaccination before allowing foreigners within their borders. Vaccines are the protector of health on the inside and continue to be one of the greatest weapons in the war.

Disinfection and Antisepsis

While Jenner provided a means to prevent infection in the long term, a number of researchers worldwide were concocting ways to rapidly kill Leeuwenhoek's animalcules. Although few realized the applicability of their actions, and even fewer related them to human health, one physician used a career disappointment to change the face of the war.

In 1865, a thirty-eight-year-old Dr. Joseph Lister was dealing with the death of a dream. He had been looking forward to a new position as a prominent surgeon in Edinburgh, but learned at the last moment that he was not getting the job; he had to remain in Glasgow. To keep a stiff upper lip, he turned his focus toward something that might increase his presence in the medical community and perhaps allow him to attain a small measure of revenge for being passed over. His target was the rather high number of infections inside the hospital environment and the need to put an end to the rising levels of infection and death.

Lister had become acquainted with the work of Pasteur and wanted to use the Germ Theory to identify the cause of several diseases that originated in the hospital. He started by investigating wounds to find whether organisms were indeed inside them. He found that there was a similarity between the germs found in different wounds, suggesting that the germs themselves were being transferred. As patients were usually too weak to move around to spread the disease, Lister had no other choice but to suspect the hospital staff. Lister came to the realization that his one source of infection was actually several hundred sources, and that each staff member was potentially a carrier of disease.

Lister tried to find the best means of stopping an infection that came from different sources and came upon the work of a fellow Scot, Dr. Robert Angus Smith, who worked on the disinfection of sewage. By using the chemical carbolic acid (also known as phenol), Smith was able to stop the transfer of germs from the different households to the environment. Taking a similar tack, Lister asked that each staff member wash their hands with a dilute solution of the acid. The intervention worked, and levels of infection dropped. Lister was soon hailed as the father of antisepsis, and his work was regarded as the beginnings of infection control. Today, there are thousands of different types of disinfectants and antiseptics in the market using agents such as alcohol, chlorine and hydrogen peroxide, all of which have shown their ability to eliminate germs and make our hands and surfaces safe.

Antibiotics

Weapons are usually the product of deliberate deduction and calculation requiring years of theory, design and field-testing. In the world of guns, swords and bombs, there are no accidental discoveries. But in the war on germs, serendipity can lead to victory. On the morning of September 28, 1928, a failed experiment led to the final great weapon.

Alexander Fleming was a proud Scot who wanted to serve his country. In 1900, at the age of twenty, he enrolled in the citizen army known as the Volunteer Force and learned of the various career paths open to him. He chose to follow the footsteps of his elder brother Tom, and became a physician six years later. During World War I, his skills made him an excellent choice for captain in the Royal Army Medical Corps, and he served commendably in numerous field hospitals. There, he faced the true scourge of infection and witnessed countless deaths. He vowed that, once the war was over, he would focus the rest of his career on finding new weapons to fight germs and prevent the horrendous consequences he had witnessed.

His work was not monumental at first, but it did offer some perspective on how germs attack the body and how internal infections were the hardest to combat. None of the disinfectants and antiseptics in use at the time was useful inside the body; quite simply, they were too toxic. Fleming spent the next decade learning more about the germs that led to internal infection, but never found a solution. He fell into the usual routine of a researcher, working long hours to find an answer, knowing that one might never come.

At the beginning of September 1928, Fleming needed to take a vacation but wanted to be sure he could start fresh upon his return. He followed the standard protocol of starting a number of cultures on petri plates filled with a solid nutrient agar, put them aside to grow, and left to enjoy his time away from the lab. Left alone with a happy environment and at least a two-month supply of nutrients, the bacteria were expected to thrive into the trillions. Upon his return on September 28, however, Fleming discovered that he had made a mistake common among novices: he had contaminated his cultures.

In order to best study a particular microbe, a pure culture has to be made. This entails taking a number of bacteria from the same species and placing them into a sterile environment, usually a broth or a petri plate. This is performed using a procedure known as *aseptic technique* to prevent any other germs from entering the environment and also growing. This procedure is conducted in a small space where airflow is usually kept to a minimum, a flame is kept nearby to keep the air safe, and instruments are dipped in alcohol followed by flaming to ensure their sterility. The technique takes practice and needs a determined focus to ensure sterility. For someone who might be distracted from their focus by, say, an impending vacation, there is a good chance that asepsis will be compromised and the cultures will end up contaminated. The experiment is lost and the researcher must undergo the following procedure:

- *Curse.*
- *Wonder how he or she could have been so stupid.*
- *Throw everything into a bleach solution.*
- *Start again, this time with better focus.*

But Fleming did something else, and took a closer look at the contaminating colonies to see if he could learn anything. He noticed that on one of his plates, a fungus had somehow prevented his intended bacterium from growing in the surrounding area. At first, the observation was depressing, because he knew his culture was lost; but after a few moments of clear deduction, his mood—and his career prospects—changed forever.

He had just discovered a new weapon.

Upon further examination of the fungus, Fleming learned that it was a *Penicillium* species. He spent the next decade learning the properties of the fungus and identified the chemical constituent that killed bacteria. As with the discoverer of *T. aquaticus*, Thomas Brock, Fleming did not choose a self-congratulatory name—perhaps because "Flemingin" had no ring to it. Rather, he called the chemical "penicillin."

In 1941, penicillin was deemed to be safe for human consumption, and within a year, clinical trials were underway, yielding nothing but positive results. In short order, the product was produced en masse and quickly became the newest and most popular weapon on the planet. The age of antibiotics had begun, and soon germs everywhere had to contend with a more powerful and ubiquitous opponent. Antibiotics continue to be the go-to treatment for any bacterial, fungal, protozoan or worm infection, and in the majority of cases, they can stem or even halt an outbreak, saving potentially thousands of lives.

A Concerted Effort

By the end of the Second World War, many powerful weapons had been developed for combatting germs, but an international effort towards their coordinated use seemed practically impossible. Attempts had been made in the past, but with little success. In 1867, the International Committee of the Red Cross was formed as a humanitarian branch of the Geneva Convention. However, the focus on germs was never on the table; other relief efforts were understandably higher on the priority scale. The League of Nations set up its Health Organization in 1924, but owing to bureaucracy, a lack of funds and conflicting ideologies, it failed and disappeared. Then, in April of 1945, during a meeting in San Francisco that led to the development of the United Nations, three doctors who sat down for lunch managed to find success where so many had failed before.

While global leaders talked of war prevention, the three physicians—Dr. Szeming Sze of China, Dr. Geraldo de Paula Souza of Brazil and Dr. Karl Evang from Norway—convened in a lunchroom and discussed their less-than-stellar roles at the conference. Evang was by far the least content. In his home country, he had become a bit of a celebrity, with his own radio program and continual coverage in the newspapers. The general consensus was that one should never debate him unless they wished to lose. He decided that he and the others needed to make an impact and came up with the idea of a worldwide health-based organization that mimicked his work in Norway. Szeming and Souza agreed.

Whether it was Evang's reputation or the actual sensibility of the suggestion, the organizers agreed, and six months later, a meeting was held to create the World Health Organization. Within a year, the WHO was in place and had a mandate to provide global citizens the opportunity to "attain the highest possible level of health." Dr. Brock Chisholm, a Canadian physician-bureaucrat, was nominated as the first director general, and he wasted no time identifying germs as World Enemy Number 1. Other newly formed health organizations soon joined in, including the Pan American Health Organization (PAHO), the United Nations International Children's Emergency Fund (UNICEF) and the United States Centers for Disease Control and Prevention (CDC). A widespread community of health advocates and workers became mobilized in an unprecedented fashion to use the weapons to fight infections, end their spread and prevent them from occurring in the future.

The work was a success, reducing the level of disease worldwide in a matter of a decade. Victory was deemed achievable, and the vision was theoretically proven a decade later, when the WHO announced in 1979 that it had successfully eradicated smallpox thanks to widespread vaccination campaigns and the cessation of outbreaks when they appeared. In the eyes of the world's health leaders, it would only be a matter of time before all infections were eradicated and microbial disease became nothing more than a distant memory of history and the world.

Beware Lengthy Campaigns

The removal of smallpox from the planet has left the blueprint to victory against other well-known targets, including polio, tuberculosis and cholera. The health community has learned exactly how germs infect and spread, and how to minimize their impact through good living conditions and hygiene. Moreover, should an infectious attack happen, the weapons are easy to use and can quickly be disseminated widely across the globe thanks to the consolidation of forces.

By all accounts, the war is ours to lose. However, germs are far from being counted out. Since Stewart's alleged announcement of victory against them, the germ code of evolution and adaptation has come into full effect. Antimicrobial resistance, which at first was isolated to labs and a small group of hospitals, has since spread into the community and eventually worldwide. Newly evolved infections have been discovered, forcing medical professionals to spend countless days and years identifying the responsible microbe using Koch's postulates, and only then learning how to fight it. An increase in the transfer of microbes from animal or insect to humans, similar to Sarah Nelmes's cowpox infection but with more drastic effects, has led to the realization that animal health was just as important as human health. And the conquest to eradicate another disease—in this case, polio—has been a bitter failure because of its ability to hide away in the environment and then return when humans are looking (or vaccinating) the other way.

But the germ code is not the only contributing factor to the prolongation of this war; human nature continues to play a starring role. Our population is growing and becoming more urbanized, while our reliance on agriculture is unprecedented, leaving us to face the same problems as the Egyptians. We continue to see our water reserves get smaller, and sanitation continues to be a problem over most of the world, suggesting that the plight of the Romans may return in some areas of the world. Human travel continues to increase, enabling germs to hitch rides from endemic areas to other regions, allowing for Hippocrates's warnings to come to light. But worst of all, our once-decisive weapons, antibiotics, while still effective, are losing ground as they are misused by those who care less about the future and only about today.

In short, while the germ code will keep pathogens in the fight, human nature might serve as their best ally in making sure the war will never end.

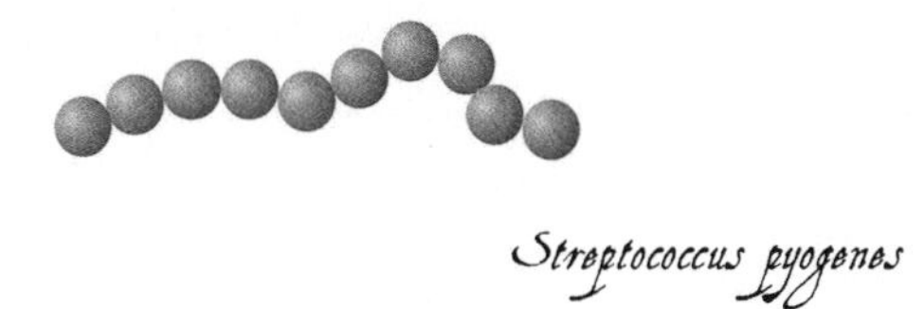

3.

EVERY CROWD HAS A GERMY LINING

For those of us who don't live in the desert, or on a deserted island, crowds are an unavoidable part of human life. Some crowds we seek out—to celebrate or cheer on our team or to protest the government en masse—and some we have thrust upon us, by department store sales and public transport. Every crowd is, by definition, a shared experience. And something shared among every crowd is germs. With every handshake, every hug, every kiss—and, inevitably, every breath—germs are circulating from one person to the next. The process is unseen, but the potential for spread can be measured through a simple experiment that can be performed without the need for ethical review board approval or lab space.

First, count the number of people within three metres around you. In a dense crowd, this could be up to fifty people. Next, observe how many of them cough, sneeze or generally look ill. Divide the number by the number of people you counted initially. This fraction shows just how prevalent illness happens to be in your area and offers some indication of the likelihood that the infection may come to you. If the ratio is less than one-third, then the risk of infection is low. If it starts getting close to one, then a visit to the drugstore might be a good idea—tissues and over-the-counter medications may soon be in order.

This experiment is the most basic form of epidemiology and provides critical information that can be used to help us in our fight. *Epidemiology* means the study of that which "resides within" a population, as opposed to what's visited upon it—a distinction made by Hippocrates in his work on understanding and controlling the spread of health threats such as infectious diseases. The practice requires the use of the kind of complex mathematical and statistical equations many of us feared in high school and college or university. But while the work may seem tedious and painstaking—and my epidemiologist colleagues assure me, it really is—its findings have been critical to our understanding of the war on germs. We have learned the various ways infections manage to find their way into a person to cause infection and then transfer it to another; how germs tend to act and react to the various efforts to control them; and most important, how human social behaviour is allowing germs to win the war and what means we can use to fight back successfully and avoid defeat.

THE ONE BODY

For most germs, the human terrain is nothing more than a nice place to visit, and one not worth staying. Ecologically speaking, these *commensal* germs rely on the earth itself and could not care less if humans—or, for that matter, any other forms of animal life—are sharing their space. There is no better example of this principle than the ever-present bacterium *Deinococcus radiodurans,* with its ability to live quite happily anywhere on this planet—and possibly others. A smaller group, *mutualistic* germs, actually enjoys living on or in the human terrain and want to keep those bodies as accommodating as possible. From an ecological perspective, these germs love to colonize on or in the body, and in return they offer either protection or regulation of the body's functions. Many, known as *symbiotic* germs, comprise the microflora on the skin as well as in the respiratory and gastrointestinal tracts; they offer assistance when there is a potential pathogen present by sending out toxins and other chemicals designed to hurt and/or kill these offenders.

Amongst these, however, are some who turn out to be turncoats. We call them *opportunistic* germs, and they are normally mutual benefactors, but if given the right opportunity, they cannot help themselves and start to attack their home environment. The most prevalent such germ of infection is *Staphylococcus aureus*. Normally, this germ is a resident who helps to keep real pathogens at bay. But in certain circumstances, the microbe turns into a pathogen itself, acquiring a taste for flesh and other organ tissues. Infection can lead to skin problems—including the never-healing skin rash impetigo—gastrointestinal distress through overproduction of toxins, and, if severe, pneumonia or other life-threatening conditions, including toxic shock syndrome.

The human terrain for commensals, mutuals and opportunistic germs is not a significant concern for us; these microbes have generally learned to live within their territories and avoid creating problems. For pathogens, however, the human terrain is a land to be conquered through infection, and each microbial species has developed ingenious and at times malicious strategies for battle.

PORTALS OF ENTRY AND PATHOGENESIS

At one time, infections were believed to lie dormant inside the body until the body underwent stress. Pasteur's revelations on germs and the development of the Germ Theory dispelled the myth, proving infections had to come from somewhere. In a corporeal sense, infection can only arise if the pathogen finds its way into the body. While the body for the most part is resistant to infections due to the natural skin barrier, there are several circumstances under which the skin can be bypassed, allowing a portal of entry for the pathogens to the body.

The major portals of entry are better known as the "Five Is," and they refer to the action allowing entry rather than the site itself. They are:

- *Inhalation*
- *Ingestion*
- *Injury*
- *Ingress*
- *Intercourse*

They are pretty self-explanatory, but in the context of epidemiology, each one stands as its own research focus, with parameters not only on the route, but also on the consequences of infection via that route, the potential consequences associated with infection, and the best means of combatting the problem using the weapons of war or other holistic options.

Inhalation

Breathing is the most important function in human life. The body can normally only survive a few minutes without oxygen before severe consequences, including brain damage, organ failure and death, occur. The lungs are usually a sterile environment, but colonization and subsequent infections do occur. Upon the appearance of a foreign organism, the immune system seeks out the invader and kills it before infection can occur. This cascade involves several cell types that have a specific function to ensure the best possible outcome. A number of pathogens cause run-of-the-mill infections such as head colds, bronchitis and pneumonia, and in the majority of cases, these are temporary; eventually, the immune system succeeds in winning the battle. But some can be rather dangerous and require a regimen of antibiotics or antivirals to ensure recovery. Unfortunately, in many parts of the globe where these weapons are not available, the consequences are dire and an infection can be life-threatening. Respiratory illnesses continue to be one of the leading causes of death worldwide.

Most respiratory infections are acute, and the situation is resolved quickly. In the case of patients suffering from cystic fibrosis, however, the fight against infection is a lifelong struggle. The cause is a defective protein known as the cystic fibrosis transmembrane conductance regulator, or CFTR. This protein is normally responsible for pushing certain chemicals, including sodium bicarbonate—also known as baking soda—into the lungs to help produce a thin, even layer of mucus. This coating acts to protect cells from contact with germs and also sparks an immune response should a pathogen make contact. In CF patients, however, the dysfunctional CFTR gene causes a change in this mucus, making it thick and uneven, leaving many cells vulnerable to infection.

The ubiquitous bacterium *Pseudomonas aeruginosa* usually poses no harm to healthy individuals, but in CF patients, this germ has evolved to take full advantage of their weakness and establishes a happy residence on unprotected cells. Once there, these bacteria have evolved to create their own thick mucus layer, protecting them against the body's defences. The mucus also acts as a foundation to allow the formation of a three-dimensional structure known as a biofilm over the cells, allowing for millions of bacteria to live in a microscopic metropolis. The consequences include a lack of oxygen at the surface level of the lungs, causing shortness of breath, a continual and maddening cough as a result of the irritation of the mucus, and recurrences of secondary pneumonia where bacteria or viruses find a home in the biofilm matrix and grow freely, posing a severe risk to health and life.

Normally, a *Pseudomonas* infection can be combatted with a regimen of antibiotics. However, in many cases of CF, the antibiotics are simply unable to reach the bacteria because of the protective layer of heavy mucus. Even more frustrating is the fact that bacteria can develop resistance to these antibiotics by using the germ code to devise a mechanism to prevent harm. In the presence of multiple-drug resistance, more holistic measures may be undertaken, including bronchial lavages, immunotherapy, and a new one with a promising future. Recently, a group of researchers at the University of Iowa tested whether or not inhaling a diluted baking soda solution might offer help. It did in pigs, and soon, they hope, it will in humans as well.

Ingestion

In addition to breathing, humans need to eat and drink to stay alive, and a number of pathogens have exploited this need in order to gain access to the human gastrointestinal system. Amongst them are the *noroviruses,* which, although not lethal, are perhaps the most egregious gastrointestinal disease. At only 30 billionths of a metre in size, this virus has shown that watching what you eat is more than just an anachronistic word of advice, it's a way of preventing seventy-two hours of hell.

Human noroviruses were not even known until 1972, when Albert Kapikian, a researcher at the U.S. National Institute of Allergy and Infectious Diseases, looked for the cause of an outbreak of diarrhea in Norwalk, Ohio. The virus had met all of Koch's principles, thanks to another one of those ethical board–nightmare experiments in which Kapikian's colleague, Robert Chanock, cultured the diarrhea and, after filtering to get rid of the chunks, gave it to a group of volunteers as an experimental drink. They became sick as expected, but Chanock wasn't done. He collected the stool from these victims and repeated the process on another group of unsuspecting volunteers, again with similar results. This process, known as *passaging,* proved that the virus could be spread and cause infections.

Kapikian took the stool samples from the original victims as well as the volunteers and looked at them using an electron microscope. He found a high level of a small spherical virus he'd never seen before. He called the new strain "Norwalk," but virus taxonomists—a rather fastidious group—decided a more general term was needed. After engaging in a deliberation filled with verbal diarrhea, they decided on the rather uninspired name "norovirus."

When a norovirus enters the body, it immediately finds a home in the lining of the small intestine. Using a method similar to a lock-and-key mechanism, it attaches to the resident cells and forces them to gulp the virus inside, a process known as *invagination.* Once the virus gets inside the cell, it starts a staggeringly fast process of replication that makes *E. coli*'s twenty-minute generation seem slow. Within six to eight hours, the virus can increase from a mere hundred particles to more than 10 billion per millilitre. Not surprisingly, the cells are obliterated and the gut becomes filled with fluid and waste, forcing a rush of severe diarrhea. Reacting to this shock, the intestines send a signal to the brain that there is an enemy inside and it needs to be expelled as quickly as possible. The easiest way to accomplish this is through projectile vomiting. As a result, the victim literally explodes from both ends and is left having to stay close to a toilet for upwards of three days. Dehydration is common, and many need to be hospitalized to receive fluids—which thankfully are kept in the body—intravenously.

Once the virus has all but eliminated all the cells inside the gut, the symptoms subside and the infection stops nearly as quickly as it started. But the experience is never forgotten.

Noroviruses have been found in almost every food source, from fresh vegetables to processed foods to water, and in the last forty years have become the most prevalent gastrointestinal illnesses worldwide. The only means of preventing infection is through the ingestion of safe, clean food and water, although other means of spreading have been documented. In 2012, a norovirus outbreak among a girls' soccer team stemmed from a contaminated reusable bag that had been in the same bathroom as an unfortunate norovirus victim. The bag was then used, without washing, to hold cookies consumed at a gathering later in the day. The food inside the bag was safe, but the virus had survived in high enough numbers to contaminate the cookies and subsequently infect the girls. This has raised significant concern about the indirect contamination of food through contact with infected sources and raised the bar of concern—it is advisable not only to watch what you eat, but also to be aware of whatever has come into contact with what you are about to eat.

Injury

A break in the skin barrier is an open invitation to germs. While many find a comfortable commensal home on the surface, the unveiling of the warm nutrients underneath is irresistible. A simple paper cut can lead to an infection, although the likelihood is slim, especially considering today's first aid measures of cleaning the wound, disinfecting it, covering it with a healing cream containing an antibiotic and appropriate bandaging. The majority of wound infections stem from much larger breaks in the skin, including severe burns and traumatic wounds. Most are treatable, but in some instances, a group of hungry germs can develop a condition whose name is almost as frightening as its symptoms.

Necrotizing fasciitis is better known as flesh-eating disease, and for good reason. The infection starts as a result of an improperly treated wound that allows bacteria such as *Aeromonas hydrophila*, *Clostridium perfringens* and *Streptococcus pyogenes* to spread from the site of the wound to the rest of the body. These bacteria literally rip open the body's own skin and connective tissue—the fascia—by releasing a number of toxins. Much-needed nutrients are freed up, and the dead matter left behind turns black, a process known as necrotization. The body, recognizing the attack, responds with an immune response to counter, but these sly soldiers have used the germ code to develop inventive measures to keep their campaign alive.

The main actor in this evasion is an enzyme known as a protease. The invading bacteria produce and then secrete the enzyme into the surrounding environment in the hunt for immunological patrolling proteins that continually look for an attack and signal back to the rest of the immune system for help. The protease finds these sentries and kills them by breaking them apart, leaving the immune cells without a signal to guide them to the intruders. As these cells have only one mission—to kill other cells—they go blindly on the rampage against whatever might be around—bacteria and host cells alike. The ravages of the immune system are swift and deadly, and within hours, entire patches of the body can be eliminated. Worse, the immune system begins to enjoy its job and initiates a spread to find other areas where they can attack. The entire body is at risk.

Flesh-eating disease is lethal without significant amputation, or debridement, which involves the scraping and cutting of infected areas to remove the infection and the dead tissue. However, with the revelations of the mechanisms behind this injury-related emergency, medicine has found ways to both attack the bacteria using antibiotics and soothe the immune system's fury using anti-inflammatory agents, including ibuprofen. Together, these treatments have helped increase the chance of life after injury, and thanks to another novel treatment—the use of high levels of oxygen in a hyperbaric chamber—the opportunity to heal and continue to live a normal life is increased.

Ingress

Unsafe injection is the predominant form of ingress and is one of the main reasons behind the spread of a number of blood-based infections, including HIV. But while these deliberate breakages of the skin are preventable, other forms of ingress are simply unavoidable. In one particular case, the body allows the invasion without the person even being aware of it, and the causative germ takes full advantage in a tragic process of events.

Warm waters are a haven for swimmers worldwide, and the only consequence is perhaps waterlogged skin. But lurking in a small selection of these waters is a protozoan species with a zombie-like appetite for the human brain. *Naegleria fowleri* is a relatively uncommon amoeba found in only a few places on the planet. But its scarcity is by far made up for in the ferocity of infection. While there are only a handful of cases each year, this parasite is a prime example of how the germ code can be used to maximize the opportunity of ingress.

N. fowleri enters the body through the nose and sets up shelter in the nasal cavity, attaching to the nasal mucous membrane. Once there, it starts to release enzymes designed specifically to break down the walls of cells, releasing nutrients for its survival and reproduction. If the amoeba were content with this, infection would be rather annoying but easily treatable. However, the germ code has given these creatures an appetite for nerve cells, and while the nasal cavity is adequate, the brain is significantly tastier. Using the olfactory nerves as a guide and appetizer, the *Naegleria* move upward from the sinuses into the brain, where they start to eat away at the matter in order to find a happy home. Meanwhile, other than some odd sensations in the nasal region, the infected person has little to no idea what is happening until it's too late. Usually, the first presentation to medical professionals happens upon the observation of delirium and seizures, indicating the brain has started to shut down. Within two weeks, the patient will die as a result of the infection spreading to the brain stem and turning off the breathing function.

This condition, known as *primary amoebic meningoencephalitis,* is almost exclusively fatal because treatment can only stop the infection from progressing; the damage done is irreversible. It's an incredibly tragic course of events no one should suffer, especially those who simply wanted to venture out into the lakes, streams, hot springs or rivers to take a nice, refreshing dip.

Intercourse

Germs have made sure to take full advantage of this fundamental requirement of human life. While talk of sexually transmitted infections has been dominated in the last few years by HIV, the lack of discussion of several other infections has allowed them to return with a vengeance, and chapter 4 is devoted to them.

TRANSMISSION AND SPREAD

If you happen to creep up on a microbiologist or epidemiologist in the hope of scaring them, the best tactic is not to use the word *boo.* Instead, say in a very soft but upsetting tone, "Ebola." The mere mention of the word forces anyone trained in the science of infection to undergo an involuntary tightening of certain posterior muscles known as the "pucker response." Moreover, the brain is forced into a spiral of horrific thoughts associated with the impact of infection on the body as well as the terrifying rate of spread in a crowd.

Ebola is a virus of about 80 billionths of a metre in length and was first discovered in 1976 in the jungles of Zaire, now the Democratic Republic of Congo. While its origins continue to confound researchers, the consequences of infection are well known. The virus enters the body through inhalation or through breaks in the skin and within forty-eight hours causes symptoms no different than the flu, marked by fever, chills and general malaise. But the virus has an uncanny ability to spread throughout the body by infecting and hiding in the cells of the immune system, the same ones given the function of stopping infection. By day five, diarrhea, vomiting and shortness of breath occur, quickly followed by blurring in the eyes and a dysfunction of the nervous system, leading to confusion and delirium. After another day or two, hemorrhagic complications begin as the body loses the ability to clot blood properly. Blood seeps through the skin and comes out of other orifices, including the ears and, most shockingly, the eyes. On the inside of the body, the blood begins to fill up the empty cavities and prevents the organs from functioning naturally. Within ten days of enduring this hell, the body finally gives up and the patient dies.

The virus can survive for weeks in the absence of a living body, and only a few viruses are needed through inhalation, ingestion or ingress in order for infection to occur. When trying to treat the living or bury the dead, a full-body suit with its own oxygen supply is needed to prevent subsequent infection. Epidemiologically speaking, a single case could theoretically lead to the infection of an entire crowd of hundreds, leading to a potentially uncontrollable outbreak.

Not surprisingly, Ebola has been the focus of documentaries, books and movies and continues to be on the agenda of health organizations worldwide. Yet, upon taking a look at Ebola's impact on humanity, the focus might not be justified. There are only a few dozen epidemics on record, and among those, many were isolated to a particular region. An infected person has never made it past their own locality to spread the disease, and if left alone, Ebola will eventually peter out as the number of victims available decreases and the virus has no place to go but back into hiding. Even in a situation where there is a fresh supply of victims, such as laboratory workers working on the virus, infection has always been limited and no outbreaks have ever been reported.

This leads to a quandary. If Ebola is such a great killer and strikes fear into the sphincters of researchers worldwide, then why is it so inadequate at taking the next step and joining the plague and smallpox as global killers? The answer lies in its transmissibility, or the actual ability of the pathogen to spread from one person to others *and* cause infection, creating what are known as secondary cases.

In epidemiological terms, transmissibility is expressed in terms of what is known as the basic reproductive number, or R_0 (pronounced "R-naught"). The number relates to the potential of an infection to spread from one person to another, called a secondary infection, and the likelihood of the development of an outbreak and its ability to turn into a pandemic.

The number is universally applicable, and some inventive—or perhaps bored—mathematicians and statisticians have used the R_0 for a number of "contagions," including the popularity of a music star such as Justin Bieber, the number of times a social media message such as a tweet is shared, or the success of an advertising campaign to get consumers to "tell two friends."

The result of the R_0 is fairly simple to interpret. If the number is between zero and one, then the infection is pretty much non-transmissible and the likelihood of spread is minimal. If the number is between one and two, there is a good chance of an outbreak and pandemic unless it is stamped out before it can spread. If the number rises to between two and five, there is a good chance the infection will spread, and unless limits are placed on the movement of the susceptible population and those who have the illness are quarantined, a pandemic is imminent. For an infection with an R_0 above five, there's little hope of stopping the spread, and the only options are to mitigate the attacks as they happen. Measles has the highest R_0 of any biological pathogen, nearing eighteen in an unvaccinated population. Without the protection of the MMR vaccine, everyone will get infected, leaving only treatment options available. The greatest R_0 calculated has been for an affliction known as "Bieber Fever." It is about twenty-four, making it the most contagious condition ever recorded.

The process of determining the R_0 is intricate and involves the use of complex calculus and statistics equations that, quite honestly, confound most people. As a result, epidemiologists have developed summary models to help explain how they arrive at these numbers. The most simplistic model is known as SIR, in which a ratio is calculated between the number of individuals who are disease-free but susceptible to infection (S), the number of people who are actually infected (I), and those who have had the infection resolved through either recovery or death (R). A higher number suggests that the germ is able to spread to those who are susceptible and maintain itself in the community through the prevention of a rapid resolution. *Yersinia pestis*, the cause of the Black Death, had an R_0 of about 3.5, making it a pathogen that needed to be controlled to prevent a pandemic. As the example in Caffa proved, a lack of appropriate containment allowed the disease to spread across Europe and the Middle East. Today, the instant that plague is discovered, the infected individuals are quarantined to prevent spread and to allow the infected to recover with the help of antibiotics. The nefarious norovirus has a similar R_0 of about 3.75, but as it cannot be treated medically, the virus must be contained through isolation of the sick and limited visits from others. On a cruise ship, for example, unless the infected people are given their own cabin and visits are curtailed, the entire ship is at risk, and many of the still-healthy passengers have had to end their voyage early as a result of an uncontrolled outbreak. At its height, smallpox had an R_0 close to eight—it was unstoppable until it had infected everyone. Yet, after the introduction of vaccination, the R_0 dropped as the number of susceptible individuals all but disappeared. As for the brain-eating *Naegleria fowleri*, infection is limited to the affected person and does not spread. Its R_0 is zero.

So what, then, of Ebola? Does this virus of nightmare proportion surpass smallpox? Does it have the ability to cause the pandemic we all fear? Or is it just a one-hit wonder that pokes up every now and then, making everyone take notice and then, just as quickly, disappearing from sight? The answer is the latter. From the outbreaks recorded, the R_0 of Ebola has been calculated to be somewhere between 1.5 and 2.0. It has pandemic potential, but as symptoms are so ferocious, the infected are barely able to move around, thus pre-empting the potential for wider spread. There is little chance for an Ebola pandemic, although the concern is ever-present.

SURVEILLANCE

With an understanding of pathogenesis and transmissibility, epidemiologists have gained valuable insight into how infections occur and how likely they are to spread. But as history has learned, outbreaks happen, and in many cases, they come out of nowhere. This has led to the development of perhaps the most valuable tool in the war on germs: reconnaissance or surveillance. The observation of outbreaks as they start and the monitoring of their progress provide the ability to spot future outbreaks and potentially stop them before they can grow.

Surveillance has been a part of public health for centuries, and initially began by enlisting the public in helping to identify potential problems. During the Middle Ages, in times of plague and smallpox, the reporting of infection was not comprehensive; the number of physicians who could treat the sick was limited, and there was no framework in place to collect and analyze records about those treated. Outbreaks and epidemics were described through the publication of letters to various societies that had little to no influence on public health. This trend continued for centuries until public health became a critical component in assuring productivity in the industrialized world.

Many attempts were made in Europe to standardize and bureaucratize health, but most failed as governments found it difficult to maintain order in the face of an outbreak. In 1866, after several nasty outbreaks of gastrointestinal infection that crippled the workforce, New York City's Metropolitan Board of Health created the world's first Sanitary Code. Within the sixty-eight pages were mandates on food safety, waste disposal, snow and ice removal and toilet-to-person ratios. In addition, there was a new requirement on the people of the city to report sick people to the Board of Health and include such details as symptoms, current condition and dwelling place. This rather extreme form of surveillance was effective at ensuring that proper records about the sick were kept—a process that was adopted in other cities—but could not stem the loss of life during an outbreak. The information simply came too late to allow health officials to find the source and curb its spread. A real-time method of following an outbreak was undoubtedly needed, although word of mouth was not the solution. The answer had come a decade earlier in London, but few were aware of the innovative approach.

Cartography—the making of maps—had been used by public health agencies to record the vital statistics of those who died. However, in 1854, a British physician showed that a map could also be used to help stop an outbreak. In the summer of that year, London was gripped in a cholera outbreak that was killing thousands without any sign of retreat.

Cholera is a gastrointestinal illness caused by the bacterium *Vibrio cholerae,* and it leads to unstoppable diarrhea. Without treatment, infected individuals lose all of their fluids and dehydrate to death. The bacterium is waterborne and can be found in almost any temperate climate, although it prefers warmer temperatures such as those in southern Asia. Cholera had found its way to Europe and North America during the colonization and takeover of India in the latter part of the eighteenth century. In London, outbreaks first appeared in 1832, although there were only a few dozen cases. In 1854, however, the bacterium returned, taking hundreds of lives in a few weeks.

After one incredibly bad day on which five hundred people died, John Snow, a member of the Royal College of Physicians, went out in search of a source. He discussed the outbreak with residents and asked where they might be collecting their water for home use. Snow confirmed that everyone affected had used the same pump, located on Broad Street in Soho. If he had lived in modern times, Snow would have had to prove cholera was present by collecting water samples, culturing them and then microscopically confirming its presence. However, all he had was a hunch and a prayer, and he used these to his advantage by asking the local parish and governing council to at least temporarily remove the handle from the pump. As desperate times call for desperate measures, they agreed, and the outbreak stopped almost immediately. He had found the source, and by preventing it from spreading, he had also stopped the outbreak.

A few months later, Snow was asked to justify his reasoning to the Epidemiological Society of London. He had been given time to think about his actions and realized that he had inadvertently developed a new means of tracking and even controlling an epidemic. He showed a map of the city and noted where the infected and dying lived. He pointed out that all the cases had occurred in and around Broad Street. He had looked for another possible source, but the visual evidence was conclusive: the pump had been involved. He postulated that this simple deduction of geography not only proved his hunch to be correct, but also could be used in the future to lessen the impact of future outbreaks. The Society agreed with him that mapping might be an idea worth investigating. Yet another fifty years would pass before the technique was fully adopted as a means of surveillance.

Snow's efforts were fundamental in changing the way active surveillance is done today. Thanks to technological advancements, computerized geographical information systems are now in place to identify the diagnosis of an infection as it happens and identify whether or not infections are localized or widespread. The development of "hot zones" became the norm, and in the case of certain annual infections such as influenza, the use of surveillance can accurately predict whether people are at risk of catching the flu. Global entities have since arisen with the sole intent of identifying outbreaks as they start and progress, in the hope of informing the world that a risk has been spotted. The Canadian Global Public Health Information Network (GPHIN) acts as a world hub for disseminating information to governments to help them prepare for a germy attack. Other organizations, such as ProMed and HealthMap, offer online information to the public to advise the global population about outbreaks and their progression.

The use of active surveillance also provides clues about aspects of infection such as seasonality and associations with climate, thereby providing added understanding of the relationship between germs and the environment. For example, epidemiologists can now say for certain that infections once thought to be tropical, such as dengue fever and malaria, are continuing to move northward towards Europe and North America, and the migration is associated with the effects of a warming planet. Indeed, while climate change continues to be debated on an ideological level, the movement of germs reveals that it most certainly does exist.

Surveillance in Real Time

While modern epidemiology has allowed for the quick identification of an outbreak as it happens, there is still a problem associated with a crowd-related outbreak. When a number of people gather, the potential exists for one infected person to share his or her germs with hundreds, if not thousands, of others and start an epidemic in a matter of hours or days. When such an event occurs, modern surveillance is rendered useless simply because of the time required not only to record the illness but also to determine what germ caused the problem. Even in this situation, a few days may be considered much too long to prevent further infection.

During the 1980s, there were several crowd-related outbreaks in the United States, Britain and Australia, leading epidemiologists to rethink how surveillance might be performed faster so the public can be informed in a matter of hours rather than days. They needed to have a better system specific to crowds of more than a thousand people, referred to as "mass gatherings." In order to identify the best options, they needed to observe a regularly scheduled mass gathering with a track record of recurrent infectious outbreaks.

There was no better choice than the Hajj.

In the Muslim faith, individuals must at least once in their lifetime make a pilgrimage to the city of Mecca in Saudi Arabia to take part in a collection of ceremonies commemorating the life of the prophet Muhammad. Each year, up to 2.5 million Muslims make the trek and become immersed in the at times solemn and at other times joyous atmosphere. There are, however, outbreaks, and many have been recorded, including those of respiratory diseases, food poisoning, meningitis and yellow fever.

After observing the Hajj over several years, researchers concluded that the number of infected people was always small in comparison to the number of people attending, and in many cases it represented a single source of infection that spread to a localized area within the crowd. They also learned there was little justification for the collection of sputum or blood samples, since the symptoms alone could easily identify a presumptive cause for effective means of prevention and control. While they preferred to identify the infecting germs, it was not necessary. The researchers also found that working passively to increase the number of health-care centres and primary care staff always led to a drop in infections; for some reason, the mere presence of health facilities improved people's conditions. Moreover, by advertising the message of proper hygiene and vaccination both before and during the event, the number of outbreaks dropped even further.

The experience with passive surveillance and action at the Hajj offered a new direction for the surveillance of mass gatherings and changed the perspective on how the war on germs could be won in these situations. By focusing not on the germ itself but on the symptoms and syndromes they cause, surveillance could be performed effectively in real time, allowing for care, prevention and control measures to be offered when needed. Diagnosis would inevitably happen to ensure proper record keeping and statistical analysis at a later date, but those requirements would not be a prerequisite for immediate intervention.

This new concept, known as *syndromic surveillance,* has been hailed and adopted by the World Health Organization and appears to be working beyond expectations. The strategy was the primary tactic used at the Olympic Games in London in the summer of 2012 and was found to be a resounding success. Public health officials and volunteers helped to identify anyone arriving and displaying signs of illness. The officials' mere presence was recognized by the patrons, who appreciated that they were never going to be alone should any germy wars arise. In addition to the surveillance, these workers helped to keep the message of proper hygienic practices alive amongst the crowds, promoting handwashing, food safety and covering up coughs and sneezes with elbows. Because many of the people on the ground were volunteers designated to simply observe and document symptoms rather than actually diagnose illness, additional funds were available to improve infrastructure by adding more toilets and handwashing stations. All of these efforts helped not just to keep infections at normal levels but to decrease them. The Games were the most germ-free thus far and offered a template for event organizers in the future.

Crowd Management

The Olympics proved that a mass gathering doesn't necessarily have to spread germs, although even the Games were somewhat tainted. There were still about a half-dozen cases of infection that forced athletes to either lose or withdraw from competition, and reports of illness amongst the crowds were recorded. Despite all the surveillance measures, there will inevitably be people who are going to be fighting off germs, and this reality mandates that each and every person be his or her own epidemiologist. As syndromic surveillance has shown, there's no need to know the pathogen or the R_0 in order to stop disease; surveillance and action can be carried out on a personal level. All that is needed is a good sightline and the simple mathematical equation of infected people divided by total people to know whether or not a situation might pose a risk. If it is, rather than bringing out the blood-collection vials and stool cups, it might be best simply to think twice about staying in place and look for a less infectious place in the crowd to hang out.

4.

GERMS LOVE LOVERS

Sexual intercourse is one of the most common routes for the spread of about a dozen or so germs. Each of these STIs—sexually transmitted infections—threatens human health in a unique way and, depending on the infection, can cause deleterious effects on quality of life.

These bacteria, viruses, protozoa and fungi have not only used the germ code to enjoy the human nether regions, they have also, in many cases, found ways to gain a foothold in different areas of the body, including the bloodstream all the way to the brain. Many, such as the hepatitis B virus and the human papillomavirus, can be handled through vaccination, while others, including yeast and the bacteria that cause vaginosis, can be prevented through regular cleanliness and care. However, a smaller group of these pathogens continue to spread just as easily as they did in ancient times, proving there is little chance they will ever be stopped.

Unlike other pathogens mentioned in this book, these germs have not used the germ code to any great extent; instead, they have taken advantage of the helping hand—or, in this case, genitalia—associated with sexual activity to spread and cause a nightmare for epidemiologists and public health authorities.

EARLY STIs

The ancient Egyptians were well aware of their sexual health; there are records of STIs in the papyrus scripts and in the scars on the bones of the dead. But STIs were sparse. In Greece, the opposite was true. Hippocrates pointed out to the masses that human behaviour is directly linked to disease, and that only through abstinence or the use of protective measures, including contraception made from animal hides, could infection be prevented. Despite his warnings and rather morbid descriptions of certain infections, there was no stopping the spread. Centuries later, the Indians, known for their sexual prowess as demonstrated by the *Kama Sutra*, emphasized the need for cleanliness. But infections still thrived.

To deal with this problem, in the fifth century, Buddhadasa, king of what is now Sri Lanka as well as a physician, composed a medical compendium to stem the tide of infections. His *Sarartha Sangrahaya* outlined not only the diagnosis of STIs, but also the preparation of drugs to treat the disease and preventative measures to keep the infection from spreading. But his efforts were in vain. Indeed, no culture or doctor-king could stop unsafe sexual practices and the consequential infections, although no significant outbreak or epidemic was ever recorded. That sort of affliction didn't occur until a millennium after King Buddhadasa wrote his text, when an act of war led to widespread sexual activity and a worldwide STI epidemic that has yet to be stamped out completely.

THE SCOURGE OF SYPHILIS

In 1489, Pope Innocent VIII gave King Charles VIII of France the kingdom of Naples, which at the time represented the southern half of what is now modern Italy. King Charles was thankful for the offer, although he sat in absentia for several years and allowed his place on the throne to be taken by King Ferdinand I. In 1494, after Ferdinand died, Charles felt that his territory was threatened by the successor to the throne, Alfonso II. Charles decided it was best to teach the young upstart a lesson and sent a group of twenty-five thousand men, mainly mercenaries, down to Naples to regain control of the kingdom. The war was short-lived, and victory was easy for Charles, whose troops outnumbered Alfonso's army by nearly twenty to one. Afterwards, the mercenaries were set free to celebrate their joyous triumph—and, unbeknownst to them, share their germs.

By the summer of 1495, Italian doctors were reporting an outbreak of severe skin and bone malformations in men and women who presented with numerous pustules all over the face and body; in some cases, patients had also lost their noses or toes. Physicians searched for a cause, but came up empty—although they did identify the cause as sexual contact. Unfortunately, there was little they could do to stop the spread of the malady, as the mercenaries had migrated all over Europe. Within a year, the infected were filling the hospitals in major capital cities across the continent. By the turn of the century, all of Europe was afflicted by the disease. In 1530, the disease was given a more concrete name, *syphilis,* supposedly after the first victim of the European outbreak, a shepherd known by the name Syphilus.

The spread of the disease prompted the church to declare war on sexual activity and adopt new regulations prohibiting it. Many obeyed or were forced to comply, including prostitutes, who were made to reveal their affliction by wearing a yellow handkerchief around their necks. The spread of the disease waned and eventually disappeared from the masses. However, it continued to spread in the underworld, where church edicts were significantly less influential than the promise of a few minutes of pleasure.

Over the next few centuries, the medical community struggled to figure out the cause of this disease that initially declared itself with the formation of very painful bumps on the genitalia, called *chancres* by the French, and progressed into an assortment of symptoms from necrosis of the skin and limbs to psychological disorders. The medical literature of the nineteenth century was filled with discussions of the disease and how best to treat and prevent it. But it wasn't until 1905 that the actual causative organism was finally revealed. A German zoologist, Fritz Schaudinn, took a sample of a pustule from the vulva of an infected woman and examined it under the microscope. In addition to the regular bacteria known to live in the vaginal area, he found a high number of bacteria that, under the microscope, looked like corkscrews.

He showed this to a colleague, Fred Neufeld, who had studied under Robert Koch, the same man who developed the postulates for infectious disease, and they concluded the bacterium was new and could be suspected as being associated with the disease. They initially named it *Spirochaeta pallida* ("pale yellow-green spire") and published the data in the hope that others would search for the bacteria. Sure enough, the results were confirmed by other physicians, although the information sent back to Schaudinn suggested the bacterium appeared to be more threadlike than spiral. He offered a name change to *Treponema pallidum* ("pale yellow-green-turning thread"), and the nomenclature stuck. Studies on the bacterium have since been extensive, and the entire progression of disease has been elucidated. There are actually three steps to a syphilis infection, and they can take years to decades to occur. Without intervention from antimicrobials, however, they are eventually lethal.

In primary syphilis, the bacteria find a comfortable spot to colonize just underneath the skin and reproduce to form a large mass of bacteria, leading to chancres. As progeny are produced, they are sent out into the body to find their next home, the lymph nodes. There, they send out signals to the immune system to slow down and stop fighting. If the infection hasn't been completely eliminated in the process—which happens in some 75 per cent of cases—then the bacteria decide to take a rest and wait for a better time to continue their campaign. A few months after primary syphilis has ended, secondary syphilis occurs as the bacteria spread out across the body, causing small rashes that become pustules.

Usually, the hands and feet are the targets, but the mouth and face can also be attacked.

The secondary phase is not as painful as primary syphilis and, unless seen by a specialist, may go unnoticed and untreated, allowing tertiary syphilis to come several months to years later. This lethal stage of the infection is manifested as the bacteria spread into the bones, the organs and the nervous system. The nose, fingers, toes and even eyes may experience necrosis as the bacterium signals the immune system to attack the body's cells, allowing the virus to feed. Psychological disorders, including a reduced function for language and math as well as decreased muscular reaction time as bacteria, hitch a ride into the spinal cord and eventually the brain. Patients will also suffer stroke symptoms as their brains lose the ability to transfer blood to all areas of the organ, choking brain cells of the oxygen and nutrients needed for survival. But the most likely cause of death will be heart failure as the organ and aorta become infected and either stop, or worse, rupture.

More than five hundred years since Charles's men sent syphilis around the world, the epidemic has yet to cease, and in many parts of the world, the pathogen is endemic. The world still sees about 12 million new cases annually, although there is evidence of a slow but steady decline. According to the WHO, there is always a hope that *T. pallidum* can be eradicated, given that prevention and treatment options for all three stages of infection are widely available, but unsafe sexual practices continue to plague us.

NO BARRIER TO GONORRHEA

As early as the second century, humans were afflicted with a painful STI characterized by a thick greenish-yellow fluid in the urinal tract. This condition was known as *gonorrhea,* meaning "flow of seed." Often, the symptom accompanied a syphilis infection, and many physicians at the time believed syphilis and gonorrhea were one and the same. However, in the late 1700s, several physicians noted that many patients with syphilis did not have gonorrhea symptoms, signifying another microbial cause. In 1879, a German researcher named Albert Ludwig Sigesmund Neisser decided to look at the purulent substance under the microscope in the same manner as Robert Koch. Instead of finding spiral bacteria, he found small, round bacteria that tended to group in twos; he called them *gonococci*. He published his findings, and in 1885 the name *Neisseria* was first published as a generic name for these bacteria, with the species being given the orthographically challenging name *gonorrhoeae*.

N. gonorrhoeae affects women and men differently. Men tend to present symptoms of infection within a few days of acquiring the bacteria. The immune system reacts by sending its troops to the site of infection. The result is a collection of white immune cells similar to pustules from a skin infection or phlegm in the lungs during a respiratory battle. Yet gonorrhea tends to reside in the male ureter, a rather small and limited space. The white cells have little place to go and cause a painful buildup. In women, there are rarely symptoms after primary infection, a significant reason why the bacterium can spread so well through prostitution. As the bacteria migrate up to the uterus, fallopian tubes and ovaries, the immune system finally reacts and causes inflammation throughout the entire pelvic region. Fever, irregular menstruation and painful urination may result, but unlike infection in men, the pain can be manageable to the extent that the problem may not be properly diagnosed or treated.

For both men and women, the symptoms tend to subside after a few days to weeks, but the bacteria are rarely cleared from the body. Without proper treatment with antibiotics, the infection can become cyclic as the bacteria migrate farther into the body, sparking yet another battle with the immune system. After months to years, the bacteria can enter the bloodstream and cause problems such as heart trouble and meningitis.

The spread of gonorrhea was significant during the two world wars, but took off with two social revolutions. After World War II, many countries experienced a shift towards a more urban culture to increase productivity in manufacturing, business and government activities. Rural and tribal individuals migrated to the metropolis in the hope of finding a better life. Many of these city newbies arrived from areas where traditions were different from the urban social norm. These cultures, many of which permitted extramarital affairs, resulted in an upswing of prostitution. For example, in certain areas of Africa, up to 90 per cent of gonorrhea infections were related to contact with a prostitute. Other areas where prostitution and uncontrolled promiscuity led to increases in gonorrhea included the Far East, Greenland and Australia.

Yet the numbers in these areas paled compared to the surge in cases in North America and Western Europe during a period better known as the sexual revolution. The era only lasted a generation, but it caused both social and infectious storms during the 1960s and 1970s. A changing mindset among the youth of the developed Western countries, particularly the United States, represented a backlash against religious morality and wartime efforts to keep sexual activity to a minimum. The number of premarital relationships grew, as did openness to homosexuality, albeit to a much slower extent. Pornography became more popular with the introduction of the videocassette recorder, which allowed material to be viewed in the home rather than in X-rated cinemas. But above all, the arrival of birth control gave women the right to control their bodies and enjoy the pleasure of their sexuality without worrying about an unwanted pregnancy. Unfortunately, the lack of a barrier form of contraception left women vulnerable to infection.

In 1974, a study by the Family Planning Evaluation Branch of the U.S. Centers for Disease Control and Prevention was published, revealing that the prevalence of gonorrhea in those using the Pill was twice that of those who weren't. The United States was clearly experiencing an epidemic of gonorrhea, and although not directly stated, the Pill was to blame. Either way, the study's recommendation to use barrier contraception to supplement the Pill worked, and rates soon began to drop.

Today, each year brings about 62 million infections of gonorrhea worldwide, although the numbers appear to be stabilizing. Still, the rate of infection is unacceptably high, which highlights the need for an increased focus on avoiding unprotected sex and for appropriate treatments once infected. Because of the bacteria's use of the germ code to develop and maintain antibiotic resistance (a topic that will be discussed later in this book), the options are becoming increasingly limited.

HERPES IDENTIFIED

In the mid-1960s, two new virus strains were identified as pathogens of the skin and nerves. The herpes simplex 1 viruses (HSV-1) were associated with cold sores, while herpes simplex 2 (HSV-2) was found to be the cause of genital warts and possibly cervical cancer. Both viruses cause significant annoying symptoms, including itching, burning and general malaise. In some cases, mild inflammation of the brain may occur, although the symptoms are not usually life-threatening. Symptoms last for only a few days, and eventually the symptoms and dermal eruptions disappear. The infection is not over, however: the viruses are persistent and can be a lifelong bother—up to three-quarters of those who are initially infected suffer recurrent infections. Subsequent episodes are typically less severe, but can still be a significant blight on quality of life.

In the 1970s, HSV-2 gained significant international attention as rates of the infection rose at an incredible pace. At the height of herpes awareness in the late 1970s, between 15 and 25 per cent of the sexually active population in the United States, the United Kingdom and China had suffered at least one outbreak. In comparison to the relatively low prevalence of syphilis and gonorrhea, which suggested there might be an infection issue with sexual activity, the rates of HSV-2 infection were shockingly clear. As the 1980s started, campaigns to prevent the spread of herpes were underway as public health officials hoped to stigmatize promiscuous sex and effectively put an end to the free-love generation. On an infamous *Time* magazine cover, the word HERPES was scrawled across the page in blood-red script, accompanied by the declaration "Today's Scarlet Letter" in boldface type. The article provoked the media to conduct many interviews with experts, but did little to stop the spread of infection. Most of the target audience didn't read the magazine, and those who did read the article dismissed it as another vain attempt to deny people the pleasures of sex. In an even more bizarre move, the ABC television network in the U.S. ran a television movie called *Intimate Agony,* which featured the effects of an outbreak of herpes in a resort area. The attempts were sadly unhelpful, and the rates of infection actually rose.

At the turn of the millennium, there were almost 30 per cent more cases than there had been twenty-five years earlier. The attempt to stop herpes was a complete failure. But the experience taught public health officials that the media could be an important tool in helping reduce the spread of a social disease. Even so, as the closing credits rolled on *Intimate Agony,* another virus was slowly making headlines that would draw attention to every aspect of human culture and send the entire sexual revolution into retreat.

A SEXUALLY TRANSMITTED PANDEMIC

In 1971, during the height of the sexual revolution, Arthur S. Wigfield, a very outspoken venereologist in the British city of Newcastle, wrote an opinion piece in the *British Medical Journal* in which he decried the lack of attention to STIs in an increasingly permissive society. He predicted there could be significant consequences to both health and society in general. He was all but ignored, and the rates of syphilis, gonorrhea and herpes increased.

Within a decade, a pathogen that had been circulating quietly around the globe took advantage of sexual openness and created what could be called a sexually transmitted pandemic.

In 1980, a new kind of illness was popping up in large American cities. Unlike the obvious signs of syphilis, gonorrhea or herpes, individuals were presenting symptoms of an underlying infection that could not be diagnosed. Many had chronic fatigue, accompanied by unending mild fevers, significant weight loss and continually swollen lymph nodes.

At first, doctors believed the illness was associated with a herpes-like virus known as *cytomegalovirus,* or CMV. The virus had been known for several decades as a cause of immune dysfunction. But infection was limited to young children who had very poor immune systems. The people walking through the doors of the clinics were seemingly healthy young men. Another possible cause was a different kind of herpes virus, human herpes virus 8 (HV-8), which initiated a form of cancer known as Kaposi's sarcoma. This infection was common in parts of sub-Saharan Africa as well as in patients who underwent organ transplants, but was rare in the general population. Both viruses were determined to be secondary infections as a result of a previous immune-debilitating infection. But the nature of that infection was still an enigma.

The infected individuals—at that time, all homosexual men—continued to worsen, and some eventually died, not from the weakness but from the inability to combat normally easy-to-fight infections that resulted in lethal pneumonia. The CDC was notified of the unusual deaths and reported on the cases in 1981, suggesting the cause had to be immune dysfunction as a result of an underlying infection. However, they had little idea of what the actual pathogen could be. By the end of the year, more than two hundred cases were reported, and while the CDC was mostly silent about its investigations, the media took on the role of makeshift epidemiologist, calling the agent a "gay plague." For the next few years, the term *gay-related immune deficiency* (GRID) was used to describe the sexually transmitted illness, although there was no indication that the pathogen was specifically limited to the gay population.

The name changed a year later when a new population of individuals—hemophiliacs—began to show similar symptoms and causes of death. Hemophilia is a hereditary condition that leaves individuals unable to produce clotting factors necessary to prevent bleeding. Regular transfusions are needed to keep their blood functioning properly; presumably, this was the route of infection. The United States Public Health Service Working Group on Opportunistic Infections in Hemophiliacs held a meeting in July 1982 to discuss the various investigations into the illness and realized that the cause was not limited to the gay population, but that everyone was at risk through the sharing of blood and possibly other body fluids. The group decided on a different name for the disease, replacing GRID with Acquired Immunodeficiency Syndrome—AIDS.

Following the lead of the Working Group, in 1983 the CDC provided an update on AIDS among the hemophiliac population, stating that they might have acquired the pathogen through blood transfusions and that sexual intercourse, while being the primary factor in spread, was no longer the sole means of infection. In March 1984, the CDC announced that the disease was blood-borne and that the use of intravenous drugs, blood transfusions and even mother-to-child transmission through birth were now risk factors that needed to be better monitored. A new set of precautions were released to make the public aware of the dangers of sharing blood as well as to protect health-care workers treating the sick and laboratory workers handling blood products. All of these efforts were performed without proving Koch's principles; even without a definitive cause, action was necessary to stem the spread.

In May 1984, the virus was finally found by two different groups in the United States and France, although each called it something different. The team headed by Dr. Robert Gallo at the U.S. National Institutes of Health called the virus "human T-lymphotropic virus" (HTLV-III), while their counterparts at the Institut Pasteur, led by Dr. Luc Montagnier, called the virus "lymphadenopathy-associated virus" (LAV).

What followed was an incredible two-year struggle of egos on the part of the research teams, each of which wanted to claim credit for the discovery of this modern plague and receive all the subsequent funding, awards and legacies. Eventually, the opposing sides came together and published a paper in which they finally described the virus known as the human immunodeficiency virus, known forever thereafter as HIV.

Research exploded and thousands of scientific papers were published in subsequent years. The virus was cultured, characterized and classified. The progress of infection and transmission was studied, and a clear picture of the virus's action on the human body was elucidated. The virus entered the body through blood and other bodily fluids, but rather than immediately causing an infection with symptoms, it found a hiding place in the cells of the immune system, cloaking them from any potential defence.

The virus had also figured out how to hide its genetic material in the DNA of these cells, effectively changing the ability of the cell to fight. The effect was akin to replacing the captain of a police station with a brothel madam. Over time, the cell would change its mandate from "serve and protect" to "serve and please the virus." It would subsequently stop fighting infections, choosing instead to harbour, and eventually produce more, viruses. If the cell tried to do its work of fighting off any intruder, the virus would send out proteins that would trigger a suicidal response; the protection of the body was compromised, and the body was vulnerable to other pathogens. In the same way a brothel causes a neighbourhood to depreciate as shady individuals start to lurk and eventually increase the crime rate, the body would depreciate through significant weight loss, an increasing number of infections and eventually death as a result of a lack of the ability to thrive.

What made the condition so agonizing was that it could take years—perhaps a decade—for the entire process to be complete. HIV infection was a death sentence without an execution date.

THE GLOBAL SPREAD OF AIDS

One of the epidemiologist's tasks is to identify the origin of an infection. While syphilis, gonorrhea and herpes were recognized in the bones of the ancients, deciphering the origin of HIV was a painstaking task. AIDS was only publicly known in the 1980s, though there was proof the virus had been circulating for decades before in Africa. Investigations looked back in time and found hints of AIDS-like infections over the previous three decades, particularly in Africa. In 1986, more than a thousand blood samples from African patients revealed HIV had been a problem in the Congo, and the virus may have been present in neighbouring countries.

By the turn of the millennium, theoreticians had estimated that the virus might have entered the human population in the 1930s, primarily through exposure to the blood of bush-meat primates, which are known to be infected with a similar virus, simian immunodeficiency virus (SIV). SIV would have used the germ code to become HIV and then quietly spread among humans around equatorial Africa for decades before showing up in cities in the late 1950s and early 1960s. Presumably, visitors from the West would engage in sexual activity with the infected and bring the virus back home, where they could potentially spread it to others. This theory has been validated as investigations into the history of HIV have shown the virus to have been present in the West more than ten years before the beginning of the epidemic in 1980.

A retrospective analysis of AIDS-like patients suggests that the arrival of HIV in the West from Africa was documented in Europe but not in North America (although no one had any clue at the time as to what the pathogen was). Arvid Noe was a Norwegian national who became a sailor during his teenage years in the early 1960s. He visited many corners of the world, including Africa, and medical records show that he had picked up a bacterial STI, presumably gonorrhea, along the way. When his seafaring days were over, Noe found and wed a lovely woman, and they had a daughter. He was apparently a faithful husband. But eventually, HIV took over not only his body but also those of his wife and their child. By 1976, all three were dead of AIDS.

About the same time as Noe was succumbing, HIV was making its way to Haiti in the blood of workers who had gone to work for the Congolese government. During the early to mid-1960s, thousands of Haitians spent years in Congo, and many took part in various sexual activities. When they returned home, they brought back their earnings as well as some HIV. The actual names of the infected are unknown, but the virus managed to spread undetected within the

Haitian population until it appeared in the United States more than a decade later.

While Arvid Noe and the Haitian people were limited in either their mobility or their ability to find a large number of sexual partners, the so-called "Patient Zero," Gaëtan Dugas, had ample opportunity to travel. According to the CDC, Dugas was a flight attendant who engaged in sex with men around the world. In the late '70s, he visited New York, San Francisco and Los Angeles when the gay component of the sexual revolution was hyperactive. He was a sexual partner to many men who became the first documented American cases and deaths in the AIDS epidemic.

Like the syphilitic mercenaries and the wartime sufferers of gonorrhea, the gay population unknowingly spread the virus out of the hubs and into the community. The virus made its way into the heterosexual population, into the blood supply through donations, and into the lives of people who were simply looking for a good time and not a death sentence. The CDC did what it could in the early 1980s by issuing countless statements on the need to practise abstinence or at the least safe sex, as well as condemning the use of intravenous drugs—all, apparently, to no avail. AIDS was spreading, but it wasn't something that people thought could affect them personally; rugged individualism overwhelmed common sense.

That ignorance was short-lived as the public woke up, not because of the infection and death of loved ones and relatives, but with the realization that celebrities were also vulnerable. From 1985 until the early 1990s, several announcements of celebrities becoming infected and dying of the disease filled the headlines. The actor Rock Hudson was the first A-list name, but many more soon followed, including Rudolf Nureyev, Perry Ellis, Liberace and Freddie Mercury.

The world finally took notice, but it was several years too late. The sexual revolution might have won the "war on sex" against religious and social moralists, but it had lost to a villainous virus.

SEARCHING FOR AN END TO AIDS

When faced with an apparent insurmountable challenge, empires tend to fall into a similar pattern that eventually ends in their demise. However, the sexual revolution had come too far to be beaten back down into the underground. Celebrities, many of whom had lost friends and colleagues to the disease, donated their time to extol the needs for safe sex. Movie studios also became engaged, popularizing condom use in movies ranging from mainstream comedies to pornography. Campaigns were initiated by governments to warn the public that, without knowledge of the HIV status of a potential mate, sex could mean death. This was complemented with widespread HIV testing in every major city. Slogans such as "No glove, no love" gained in popularity, and attempts were made to increase education both in the community at health-and-wellness centres and in schools, which began to provide sex education to students. While some programs insisted on the use of abstinence as the only means, most focused on ensuring that children understood, long before they became sexually active, how to play safe in the bedroom through the use of condoms and other means of pleasure that didn't involve vaginal, oral or anal intercourse.

Complementing the change in sexual behaviour, researchers have worked to find weapons to kill the virus. While treatments have been around since the mid-1980s, the virus used the germ code to develop resistance to antiviral drugs. In perhaps the most infamous case, the antiviral ziovudine (better known as AZT) was thought to be the answer to AIDS. In reality, treatments were only effective for a matter of months before resistance re-emerged. Over time, the medical community found that using a combination of treatments had the ability to reduce the level of virus in the blood, but even this option was not enough to clear the body of the virus.

In the last five years, there has been a remarkable increase in the use of novel treatment options, lessening the impact of HIV in the body and increasing the lifespan far beyond originally believed, but a cure continued to be evasive. Yet in 2012, hope that a cure was in sight was featured in the global media as reports about virus clearance and the first person ever "functionally cured" of HIV—known as the "Berlin Patient"—began to appear in the scientific literature. Soon came reports of other individuals being functionally cured through various antiviral treatments. However, whether these cases are a glimpse into the future or simply outliers in the big picture has yet to be determined. What is certain, however, is that after 30 years of frustration, there is finally a real sense of hope.

Today, the rate of infection of HIV continues to be stable at a few million new cases per year and a total of 34 million people with HIV. Efforts to control the virus, both at the molecular and social levels, are gaining ground, and there is a belief that there may be an end to the AIDS pandemic within a generation. Once AIDS has been fully eliminated, it will be a part of STI history and will be looked upon as yet another example of a time when a germ caused millions to worry about their health and lives. Whether the impact of the sexual revolution will ever be associated with the disease and its spread is an as-yet-unanswerable question. There is little doubt that promiscuity was the underlying cause, but in a world that prefers to look at sex as a means of pleasure and not of possible infection and death, history may be written to omit that particular aspect of the story, offering new and emerging STIs the opportunity to resume their assaults on humankind.

A SEXUAL SNEEZE

At the height of the public education campaign about HIV and its relationship with sex, public health officials tried to ingrain a thought into the public consciousness: when you have sex, you are not just having intercourse with another person, but with everyone in that person's sexual history. That idea was somewhat helpful in increasing awareness, but it never took hold as expected. Perhaps the analogy was missing a valuable component that could be better conveyed through a more general human activity that occurs spontaneously and could affect an entire crowd of people. Each sexual act is a potentially infectious sneeze, and a partner is susceptible to any and all expelled pathogens that have been acquired over time. The only way to prevent catching the illness is through a monogamous kinship with someone who is known to be germ-free, properly distancing oneself from known infected sneezers, or through the use of protection by both the sneezer and those in the surrounding environment.

Unfortunately, while more people tend to heed these suggestions, there will always be those who choose not to cover their sneezes or eschew an open sneezer. In the same way, those who choose to engage in the sexual sneeze openly increase the chance a germ will hitch a ride and begin an infection that could potentially lead into an outbreak or something much worse.

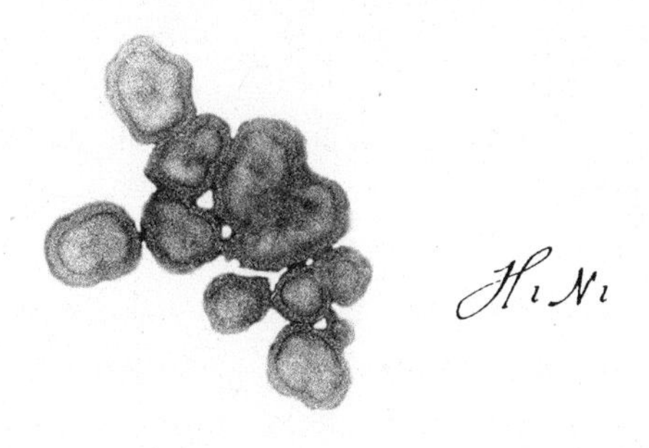

5.

THE SNEEZE THAT WENT AROUND THE WORLD

While the microbial landscape is continually evolving as germs find new and innovative ways to thrive, the true masters of evolution and adaptation are the influenza viruses. These pathogens evolve so rapidly and consistently that they maintain a yearly schedule for human infection. The "flu season" is known to every nation on the planet and annually afflicts between one and five per cent of the global population, burdening between 70 and 350 million people with three weeks of coughs, congestion, respiratory struggles and overall malaise. Of them, about 0.1 per cent will perish because of their immune systems' inability to fight off the infection.

There are some years, however, when flu turns into a real killer that leaves millions in its wake. In the last century alone, the world has been sent into a panic three times as a result of the flu's overzealous use of the germ code. In 1918, the "Spanish flu" spread across the world and killed up to 50 million people. In 1957, the "Asian flu" was less virulent, causing close to two million deaths, but still managed to gain the status of a pandemic. In 1968, the "Hong Kong flu" resulted in a million deaths and was also dubbed a pandemic. In early 2009, another flu strain made its way into pandemic history and has taught the world a number of valuable lessons on how flu can use the germ code to establish not only a seasonal presence but also the occasional pandemic one.

In March of that year, an influenza outbreak was identified in the small Mexican village of La Gloria. The only reason it even made it onto the radar of epidemiology was that nearly 50 per cent of the 1,500 inhabitants were showing up sick. A team of researchers from the World Health Organization's Pandemic Assessment Collaboration went down to Mexico to check on the virus and see whether it had pandemic potential.

The results were perplexing, at best. Normally, the majority of infections occur in the very young and the elderly, while only a small minority of cases occurs in those who have the healthiest immune systems. Yet most of the affected here were between the ages of twenty and forty-nine. The R_0 for this virus was about 1.5, which meant it had the same ability to infect as a pandemic strain, but the virus had yet to be seen outside the village. The most telling observation, however, had nothing to do with the flu but with the geography of the town.

La Gloria is located near a concentrated pig farm that holds close to a million pigs. Data from the last quarter-century has revealed that seasonal flu strains normally arise from pig farms—most notably in China, where concentrated farming had become the norm in many regions. This Mexican version of the flu seemed to be somewhat different from the ones "Made in China," but not overly concerning. The conclusion was unanimous: the potential for a pandemic was there, but the impact was going to be minimal.

At the beginning of April, shortly after the La Gloria outbreak, another outbreak of flu occurred in Mexico City. This time, however, the flu wasn't only infecting, it was killing at an incredible rate. More than 1,300 were infected, and 15 per cent of them died from the infection, which apparently went systemic and spread throughout the body. The observation stymied epidemiologists, who had all but dismissed the La Gloria strain a few months earlier as non-lethal. But while the scientists were left scratching their heads, the media were licking their lips as they jumped on the opportunity to report on this new "swine flu" that had killed in Mexico, infected people in the United States and had the potential to go global. That potential was realized in less than a month, as similar infections were reported in Canada, Spain, Israel and New Zealand. While the spread was disconcerting, the varying levels of death being recorded in different geographical areas were confounding. There appeared to be two kinds of flu spreading across the globe, and without being able to understand how the germ code applied to this phenomenon, health protection authorities were working blind.

By the end of month, the causative viruses from all over the world had been isolated and the germ code analysis was underway, but results were only preliminary and in some cases deviated from those observed in other labs in other countries. The information did lead to a narrowing down of what to name the virus—"American flu" or "Mexican flu" instead of "swine flu"—but in terms of public health decisions, there was little help for the countries dealing with death reports that rose daily. Schools were forced shut to keep children from being infected, and in some countries, overwhelmed hospitals closed their doors.

While the WHO hoped to await word on the nature of the virus and its evolution, by June 11 a total of seventy-four countries had declared the virus present in their communities and pressure grew on Director-General Margaret Chan to make the call. Despite not having all the needed information, she had no choice but to declare a pandemic, the first of the twenty-first century.

WHAT IS "SWINE FLU"?

The media popularized the name of the virus causing the pandemic, but researchers had known about "swine flu" for decades. The actual illness was first examined in 1918 in the wake of the "Spanish flu," when certain epidemiologists found that pigs and humans shared similar flu-like symptoms. Over the next decade, the flu was hunted down in other animals, and similar symptoms were again found in ferrets. In 1934, a researcher named Richard E. Shope shoved the nasal mucus from a sick pig into the nostrils of a healthy ferret and caused infection. However, the ferrets only became marginally ill compared to their porcine counterparts. He then tried to transmit the virus from one ferret to another, but found that the only way he could make infection happen was by performing the same protocol of collecting and transferring "ferret snot."

As he continued this process of passaging from one ferret to another to another, something incredible happened. With each passage of the virus, the symptoms tended to get worse. After five or six sequential transfers, the virus was able to transmit from one ferret to another naturally, through sneezes. Shope's final experiment was to attempt to reintroduce this new ferret flu to pigs, but he found that they were no longer suffering the same symptoms and recovered faster. The virus had evolved to infect ferrets and was no longer capable of infecting pigs. This trait of flu became known as *host switching* and was the basis for an entire generation of species-to-species transfer experiments.

Researchers spent most of their time examining how "swine flu" and two other human flu viruses—unimaginatively called human A influenzavirus and human B influenzavirus—infected ferrets and other animals, including horses, dogs, mice and fowl such as ducks, geese, turkeys and chicken. Shope's work was now being copied worldwide, and the art of passaging viruses from one animal to another became its own branch of research. The ability to work with lab cultures of cells (called *in vitro* work), rather than in animals (known as *in vivo* work), only increased the number of passaged strains. *In vitro* work was so much easier than *in vivo* because large amounts of the virus could be grown in as little as a few millilitres of fluid in a petri dish. These cultures could then be passaged onto a petri dish filled with new, uninfected cells and allowed to infect them. Passage time dropped from months to days, and soon, *in vitro* passaging became the norm.

Shope's pioneering work also helped to set the influenza world on fire and led to a massive expansion of research. Whereas in centuries past there was a single researcher such as Lister, Koch or Pasteur, new conglomerates of researchers began working collaboratively to share viruses between them—in vials, not in lungs—and collectively peer into the world of the flu virus. The rise in the number of researchers, students and technologists accelerated the speed of discoveries and led to scientific papers with sometimes dozens of authors.

With all the lab and animal passages, freezers began to fill up with hundreds of different strains, creating a problem for researchers. To keep track of all the flus in their freezers—and more important, in their publications—a new form of nomenclature was needed. "Swine flu" might have been a useful point of reference in the early twentieth century, but by the 1940s a new system was needed for tracking and identifying them. There were several requirements for a good nomenclature, and taxonomists worked hard to develop a descriptive model that would also be useful in the lab. The type of virus was important, as were the host species, reference code number and year of discovery. Using these points as a guide, a rather simple format was developed. For example, the most commonly used influenza virus in research is one that was first isolated in Puerto Rico in 1934 and given the lab code 8 for documentation reference. This virus is known officially as A/human/Puerto Rico/8/1934 and is still used today as a control model of flu biology.

With nomenclature established, researchers spent much of the latter part of the 1940s attempting to visualize the viruses, using the relatively new process of electron microscopy. Yet in order to prepare a sample for the microscope, it needed to be solid. By this time, flu was being mass-produced in chicken eggs and could be identified through a process of mixing red blood cells from adult chickens with blended egg fluid, forming a semi-solid goop comprised of agglutinated cells. The mass was solid enough for use in the microscope, and by 1950 the virus was finally seen. Much to their surprise, the virus didn't appear to be a monster, but something akin to Leeuwenhoek's animalcules, with an irregular round-to-oval shape. The finding was remarkable nonetheless, as researchers could finally see the virus and isolate it to enable studies at the molecular level.

The primary focus of research was to identify the molecular cause of the red blood aggregation, or hemagglutination, in chickens. With much effort, researchers finally isolated the protein responsible and gave it a fairly obvious name: hemagglutinin (H). A few years later, another very important protein was discovered; this enzyme changed the structure of a part of the lung cell, a sugar known as neuraminic acid, allowing for the virus to infect the cell. Not surprisingly, this was named neuraminidase (N). By the 1970s, sixteen distinct H proteins and nine separate N proteins had been identified, providing a final key to nomenclature in which the actual structural components of the virus can be easily classified. The most common forms of human flu were found to have the H1 and N1 as well as the H3 and N2 proteins. Thus, in addition to the classification, a new detail was added to the virus nomenclature, and the Puerto Rico strain became A/Human/Puerto Rico/8/1934 (H1N1). Today, if the virus is known to infect humans, the species component is left out such that the virus is known as A/Puerto Rico/8/1934. Yet even this is too long for most researchers; they shorten the name even further to Puerto Rico/8 or, as my colleagues call it, PR8.

By the mid-1970s, a full understanding of the influenza-virus biology and taxonomy was complete and researchers were able to focus their attention on how these little viruses evolved and created a new "line" of viruses every year. They looked inward to the monarch of the virus, the genetic material, and found something remarkable. Instead of one large piece, as found in most germs, the virus contained eight smaller pieces. These segments were encoded for different proteins, and all eight were required for a virus to survive. When researchers tried multiple infections with different strains, such as an H1N1 and an H3N2, they found that, in addition to the same viruses being created, two new ones emerged: an H1N2 and an H3N1.

The result was fascinating; it was the first time the germ code had been manifested in this manner. This mixing of H and N was novel and effective—combining two viruses from different animal populations created viruses that could infect both. This is called reassortment, or *antigenic shift,* and is the main reason the virus circulates on an annual basis. Pigs, birds and humans gather together on farms and share their viruses, allowing them to mix, reassort and find a new strain that will infect all and spread globally.

But reassortment could only do so much to explain the nature of "swine flu," how it became a pandemic and why the mortalities of the La Gloria and Mexico City versions were so different. There had to be another manifestation of the germ code to allow this kind of evolution. Researchers went back to their freezers and examined the number of viruses they had collected from passaging to see if there were any changes in the hemagglutinin protein at the amino-acid level. After all, if cold-loving bacteria could change the amino-acid structure of their enzymes to adapt to the environment, perhaps flu viruses were also making changes to accommodate the new animal environment.

The investigations proved to be fruitful, and mutations were indeed found in the H protein as well as the N and several others. In some cases, there was only a single amino-acid change, while in others, as many as four or five changes were identified. They noted significantly stronger effects, longer periods of infection, increased pathogenicity (capability of causing disease), increased transmissibility and, in some cases, increased number of human deaths.

This feature is called an *antigenic drift,* or adaptation event. Antigenic drift is fascinating even to seasoned researchers because it highlights how a minor change can lead to novel characteristics. The process of drift is similar to the minor changes in design that lead to the hottest fashion trends. The altitude of a hemline or placement of a pleat can set the fashion world on fire; the choice of zipper placement can blow the critics away. Minor structural changes can bring about major functional change. In flu biology, similar small changes in the structure of certain proteins can create an upheaval of consequences through an easier entry into cells, a higher level of pathogenicity, an increased ability to spread from the lungs to other parts of the body (including the blood, gastrointestinal tract and other organs) and a higher R_0, meaning an even better chance of transmission.

The process of antigenic drift happens each and every time a flu virus infects a cell. The virus is continually producing different versions of itself in the hope of making a better progeny. However, the success of these drifts is rare, as only one or two of these minor changes can ever really take hold. In addition, the virus has to be able to spread its evolution over more than a few people. The researchers who adapted "swine flu" to ferrets did so through forced infections, a protocol that continues to this day. The only other means by which a drift could occur is through close contact with the infected, particularly in large populations. The use of concentrated animal farms only helps to enable these drifts and offers up the chance to gain access to humans—and kill.

TRIPLE THREAT

The La Gloria strain, officially known as A/Mexico/LaGloria-3/2009 (H1N1), was found to be remarkably similar to others found worldwide, but it had made some changes that rendered it more infectious. The Mexico City strain, known as A/Mexico City/001/2009 (H1N1), had also changed—but in a different way, to make it a killer. A third strain was discovered by the CDC from an infected child in the United States and designated A/California/4/2009. The three strains were shared with researchers worldwide in the hope of identifying whether the virus had undergone shifts or drifts to spark the pandemic. The results were unanimous: the viruses had undergone both.

The genetic material of the virus was unbelievably complex. First, there were pieces from not just one or two, but three different kinds of influenza viruses, from humans to birds to pigs. This unprecedented "triple reassortant" suggested that the virus originated in Asian birds, Eurasian pigs and North American humans. What was even more incredible was that this virus took more than four decades to develop in animals—one sequence came from a 1968 strain of duck flu—and several of the viruses' ancestors had infected humans in the past and somehow returned to the animal population.

The antigenic shift was impressive. However, it was minor in comparison to the revelations of the antigenic drift. There were a number of small single-amino-acid changes in many of the proteins, but the most impressive was the change in the H protein. In this case, the change determined whether the virus would cause merely a severe infection or a lethal one. At the molecular level, the La Gloria and California change allowed the hemagglutinin protein to snuggle in more closely to the cell and increase the likelihood of infection in the lungs. The more lethal Mexico City strain, however, allowed the virus to snuggle not only with cells in the lungs, but also in the blood, causing a life-threatening condition known as viremia.

The appearance of the two significantly different mutations in the same spot with such different circumstances was a signal to researchers and epidemiologists that antigenic drift was itself dynamic and that surveillance and tracking needed to focus on identifying not just one, but perhaps many possible uses of the germ code to enact drastic consequences. The revelation also highlighted the need for dynamic response and the choice of weapons that match the severity of the infection.

A NEW WEAPON

While "swine flu" continued its global spread, the medical community was working to use its weapons to treat the sick and prevent them from dying. Two anti-influenza drugs were useful in this fight, and they were put to good use. Amantadine, which was approved for use in 1966, and oseltamivir, approved in 1999, both attach to the neuraminidase protein of flu, preventing it from entering the cell. These drugs were effective for the most part, although in less than one per cent of the virus strains found in 2009, the germ code had already been used to effect an antigenic drift, changing the structure of the neuraminidase so that one or both of the drugs could no longer attach to the enzyme and prevent infection. The antivirals were somewhat successful, but could not be wholly trusted. There had to be a more powerful weapon in the arsenal that could prevent the virus from using the germ code to keep the fight alive. That weapon was a vaccine.

The flu vaccine was first designed in the 1940s, and over time it had developed into a highly efficient protocol requiring about a year to develop and manufacture. The flu vaccine works to present exact copies of certain virus proteins to the immune system, training it to fight should an actual infection occur. Unfortunately, the vaccine is not universal, and a new one is required to deal with new strains that have undergone a shift or drift. Each year, a new "flu shot" is produced and made available worldwide to prevent infection during flu season.

As soon as the CDC had cultured the A/California/4/2009 virus, they tested previous vaccines, but none proved to be effective. Thus, a new one was required. In an unprecedented global effort, the new vaccine was ready in just a few months and ready for the public in October. The uptake was incredible, with long lines at clinics and other specially designed vaccination sites. At one point, there was concern that there might not be enough vaccine for everyone, and limits were put on the individuals who could actually get the shot. But as time wore on, the lines shortened and eventually disappeared. In the final tally, only about a third of the global population who qualified actually received the flu shot, primarily because the majority had either already been infected and/or recovered or had a poor perception of the safety and efficacy of the vaccine, considering its rather short timeline towards approval.

Perhaps the best weapon throughout the entire pandemic came not from a medical leader, but a spiritual one. In September 2009, the Dalai Lama was visiting Memphis, Tennessee, which was being hit hard by the pandemic virus—there had been close to ten thousand visits to the hospital in that month alone. The mayor of the city took the opportunity to avoid a handshake, but instead offer a fist bump to the spiritual leader. Images of the gesture were transmitted around the world, and public health officials harnessed it to demonstrate to the public that the best way to stay healthy is to avoid infection through interpersonal contact such as handshakes, hugs and kisses. Hygiene took centre stage in the news as experts and aficionados alike took to the airwaves to inform the public that handshakes, hugs, kisses and other intimate encounters were to be avoided and that people who were ill should either distance themselves from others or have that decision made for them. Schools and workplaces started to see a decrease in attendance as people stayed home to recover from their illness or simply to ensure they didn't come into contact with the virus.

While overall economic productivity may have declined, so did the numbers of infected and dying individuals. Thanks in part to the drugs and the vaccine—but also to the improved hygiene of the world's populace—the rate of infection dwindled to seasonal levels. Nations around the world were breathing more easily as 2010 began, and many declared their own victories against the virus. Yet the World Health Organization waited another eight months, until August 10, 2010, before declaring the pandemic over.

In the end, 128 countries had reported "swine flu" infections, and while lab confirmations revealed about 1.48 million infections and 25,000 deaths, estimates suggest the actual numbers might have been twenty times higher, with up to 30 million people infected and 500,000 lives lost. The global mortality rate was about 1.5 per cent, which is still high for influenza, but the local statistics show some places had it worse than others. While developed countries with rapid access to treatment and vaccines had a death rate of about one in every sixty confirmed cases, other places, such as North Korea, Hungary and Lithuania, saw mortality rates that exceeded 25 per cent.

THE AFTERMATH ANALYSIS

With the pandemic a past event, the quest to discover what happened—and, more important, why—will most likely continue for at least a decade. New insights into the day-to-day activities in the affected communities, further genetic analysis of the various influenza viruses isolated and cultured in the lab, and continued administrative reviews of the worldwide response will help explain what brought about the pandemic. While official reports and recommendations will trickle out over time, there are already observations that can be used to predict or even prevent the next pandemic.

"Swine flu" had evolved for nearly forty years in birds, pigs and humans without a single alert to its presence. This suggests that, somewhere in the seasonal flus of the past, an ancestor of this strain had found its way into humans and then somehow had returned to animals. This cyclic nature had been shown in the lab by Shope back in 1934, but never before in nature. While the information was useful for researchers, in the real world the revelation posed a significant challenge. Without proper education of farmers and the provision of suitable protective equipment to prevent transmission of disease, flu could continue to move back and forth between humans and animals, giving the virus ample opportunity to find yet another means to spread worldwide.

The impact of local evolution on the global spread of infection is also highly relevant and highlighted the need for more extensive surveillance of farms and other animal populations worldwide. The differences between the death tolls associated with the La Gloria and Mexico City strains suggest that new viruses that do not look or act alike can arise from locations close to each other. In the wake of H1N1, surveillance of swine farms in Canada and other developed countries has become a part of the agricultural regulatory system. But in many other parts of the world, surveillance continues to be poor and sporadic. These efforts need to be strengthened in order to catch the germ code in action, whether on a remote farm in the middle of China or the family-operated livestock facility just outside the city. Unfortunately, worldwide adoption of this practice is unlikely, and the potential for another pandemic strain arising from pigs is just a shift or drift away.

In addition to surveillance, there has been much attention paid to the overall health of animals. In a time when concentrated farming puts thousands, if not millions, of animals into a small area, localized evolution is to be expected. While the use of antibiotics and other antimicrobial supplements may be effective at stopping bacterial infections, none of these can stop the spread of the flu. A mass animal-vaccination campaign is also an option, but unlike for humans, who enjoy the benefit of an annual vaccine, animals have to deal with what is available at the time. This leaves many of them vulnerable to infection while increasing the chances that the virus will circulate within their population and eventually develop, through shift and drift, into a pandemic killer.

Margaret Chan, the director general of the WHO, said in 2010 as the pandemic was waning, "We got lucky." "Swine flu" had achieved pandemic status, but the impact had not been apocalyptic. Critics of the WHO tried to suggest that the lack of significant deaths in comparison to the "Spanish flu" made "swine flu" nothing more than a sheep in a wolf's clothing and should never have been taken so seriously. However, this view is short-sighted. The analysis has highlighted the needs for better surveillance, response and investigation and offered recommendations for future actions. The turn of luck should not be taken for granted; rather, it should lead to improved planning and response strategies. "Swine flu" may have been a shock, but compared to other pathogens lurking in the environment and waiting to strike, this pandemic was quite simply—and scarily—a shot across the bow.

6.

A PANDEMIC STATE OF MIND

For the officials in charge of emergency and disaster preparedness, nothing is more disconcerting than a perfect storm. The mere thought of a weather event with the right combination of damage-causing climatic conditions can send them into a state of panic, leading to unprecedented actions. In September 1999, one such perfect storm, Hurricane Floyd, encroached on the Caribbean and the East Coast of the United States. The storm was a giant, nearly 1,000 kilometres in diameter and exhibiting wind speeds in excess of 300 kilometres per hour. It was deemed the "Storm of the Century" by the media, and many in the meteorological world feared the entire East Coast could be pulverized. To prevent widespread death, more than three million people were evacuated from their homes, the largest peacetime order of its kind. When the storm hit, there was the expected significant flooding and wind damage, and many who did not heed the evacuation orders perished—fifty-seven lives were lost.

After Floyd headed north and petered away, criticisms were voiced that, while the official response to the storm had been a success, it had also been an overreaction. The damage was extensive, but not as bad as expected. Southern Florida spent millions on evacuations and solidifying buildings, yet the region wasn't even affected by the storm. Farther up the coast, New York City prepared for the worst, but only suffered marginally. The evacuations and other measures had created human costs, both economic and personal, that were deemed to be excessive. The authorities, in the eyes of many, had got it wrong.

But the problem with perfect storms is that they are unpredictable. You need to be prepared for widespread death and untold billions in losses every time. And if the storm proves to be mercifully less dreadful than predicted, you should resist calling for a less drastic protocol to deal with storms in the future.

Such was the experience with the H1N1 pandemic of 2009–10.

WAS "SWINE FLU" REALLY A PANDEMIC?

One of the greatest criticisms of the World Health Organization over its handling of the pandemic was the fact that it called the "swine flu" a pandemic in the first place. Normally, pandemics were limited to outbreaks that had a mortality rate of at least 10 per cent, which even the Mexico City outbreak couldn't match. For many countries, H1N1 appeared to be little more than an aggressive seasonal flu that didn't warrant the extra money spent on antiviral purchases and the stockpiling of a rushed vaccine. Considering that most countries were struggling in the midst of a global recession, the decisions were seen as imprudent or even foolhardy. The hundreds of millions of dollars spent on a global overreaction would have been better used elsewhere.

Yet the WHO has fought back against these claims, suggesting that, at the onset, there was a definite risk of a pandemic and that as this flu strain apparently hadn't been seen before, the effort was worthwhile. The experience also demonstrated some of the lack of animal surveillance and that the world needed to look at the health of its animals as much as that of its humans. But most of all, the year-long struggle was the perfect dry run for the future, considering that the world had been completely unprepared for one near-pandemic and refused to be serious about another that might be coming down the road.

The WHO was justified in calling the 2009 H1N1 outbreak a pandemic. It fit almost all the standard criteria, including multinational spread, high attack rates, lack of population immunity and transmissibility; the only real place where it faltered was in a high mortality rate. Also, the last worldwide outbreak had been over forty years earlier, so there was an urgent need to audit the improvements in surveillance and reaction to identify the gaps that could lead to a disaster. The H1N1 response might have been a pain for the population at large, but as many in the emergency-preparedness world will tell you, that is not too great a concern. After all, a complaining person is still a living one.

The pandemic call was also an excellent opportunity to prepare for a future that portends to be significantly different from the past. The pandemics of old, such as plague and smallpox, are no longer a threat, but they have been replaced by a plethora of worrisome new and emerging germs. Some of these have already hinted at their potential to become the next widespread killer, although they never quite made it to pandemic status. What makes these new and emerging pathogens so unique is that they have different strategies than did the spread-and-kill mechanisms of the past. Many of these enemies have taken advantage of global and human trends to cause disease and spread, while others have learned that the best way to become a pandemic pathogen is by waiting in the shadows and popping up when it is least expected. These challenges suggest there is much to learn from their examples. Their emergence may lead to illness and death, yet we can learn from them to ensure that the perfect pandemic storm never happens.

H5N1: THE LURKING PANDEMIC

One of the major criticisms of the H1N1 pandemic call was that the wrong virus was being targeted. H1N1 was not the virus everyone feared. That honour had been bestowed on another type of flu, commonly known as H5N1. This flu strain was first identified in December 1997 in Hong Kong, when some eighteen people became infected with what appeared to be normal influenza, but then several of them came down with significant respiratory complications. Six died, representing 33.3 per cent mortality.

The virus was eventually cultured and given the name A/Hong Kong/156/1997 (H5N1), which in itself was shocking—there had been no known previous cases of H5 viruses infecting humans. The H5 viruses were only known to infect avian species, and there had been no indication the strain could jump to humans. Eventually, epidemiologists hunted down the animal carriers and were horrified to learn that the source wasn't a wild animal, but a domestic one: the chicken. All their tests revealed that these birds harboured the virus and somehow transmitted infection to their human counterparts through extremely close contact, both in the markets where they were sold and in the homes where they were raised. The situation looked dire, but possibly manageable. The H5N1 virus met all the criteria for a pandemic but had not spread outside the Hong Kong area. Much like Ebola, there was an opportunity to quarantine and isolate the virus before it could spread. Unfortunately, that meant killing thousands, if not millions, of chickens, a prime source of nutrition. A meeting was held in Hong Kong to decide on a strategy, and the decision was unanimous: human safety was worth more than the economic benefits of keeping the chickens alive. A mass cull of 1.5 million chickens followed. The virus was deemed to be eliminated but not eradicated. There was little doubt it would return.

Over the next few years, surveillance for H5N1 became widespread, and epidemiologists learned that it was endemic to the entire bird population. The strain was lurking in the environment and could pop up at any moment. It would take until 2004 before the virus returned to the public eye, in

Vietnam and Thailand. Cases were sporadic, but were becoming increasingly more lethal. As the virus was spreading, it was also mutating into an even stronger killer. An isolate from a 2004 outbreak in Vietnam, denoted as A/Vietnam/1203/2004 (H5N1), revealed the virus was spreading throughout the body, fatal to anyone infected and not treated immediately with antivirals.

Surveillance measures spread like wildfire, and H5N1 was being detected in birds all across the Asian continent and into Africa, as well as in other species. H5N1 was detected in tigers, resulting in the slaughter of 140 zoo animals. It wasn't long before the virus showed up in domesticated cats. Dogs and ferrets were also showing the ability to carry the virus, although this was rare. But the worst news came with the discovery of the virus in Indonesian swine. While culling would prevent an outbreak in the country, unless every pig on the planet were slaughtered, there was no stopping the leap into humans *somewhere.* The pandemic was lying in wait, but wasn't yet ready to make its entrance into the history books.

Today, we are still waiting. H5N1 continues to make an appearance in a few countries where close contact with chickens and/or pigs is still customary, but in the rest of the world there seems to be no human infection. Research is still being done as to whether H5N1 might ever become a pandemic, but the results seem to be as evasive as the virus itself.

In the meantime, the world cannot simply ignore the situation. With an apparent near-global distribution and a continually increasing lethality, there is no questioning the need for surveillance of animals for presence of the virus—and, more important, analysis of its evolution. The clock may be ticking, but it's slow enough that we can be sure that the next time H5N1 rears its ugly hemagglutinin, we will be prepared.

THE SCOURGE OF SARS

While the majority of surveillance in late 2002 was focused on H5N1, an outbreak of a different sort was happening in China. In November of that year, a respiratory illness had infected more than eight hundred people in the Chinese province of Guangdong and killed thirty-four, some of them health-care workers. One of the physicians who helped end the outbreak, Liu Jianlu, was worn out and in need of a well-deserved break when he travelled to attend the wedding of his nephew in Hong Kong in early 2003. He arrived on February 21 and settled in at the Metropole Hotel, a nice place known for its central location and comfortable rooms. Whether it was exhaustion or the onset of a cold, Liu had not been feeling his best the week before, but he still decided to go on the trip. When he woke up the next day, he was feeling even worse and decided to seek medical attention. A few hours after arriving, he was put into intensive care, and within two weeks, he was dead.

While Liu was entering the hospital, two people he met the day before at the hotel were heading off on the next leg of their trip. Johnny Chen, a businessman from Shanghai, was heading to his next port of call in Hanoi, Vietnam, while Sui-chu Kwan, who had come all the way from Toronto with her husband for a lovely vacation visiting relatives, was heading back home to Canada.

Almost as soon as Sui-chu arrived in Toronto, she was visibly ill, and the next day, her son Chi-Kwai Tse took her to the emergency room. When she was admitted, she was suffering from respiratory distress. She never left the hospital, and died on March 5. In Vietnam, Johnny Chen started to feel what he thought were just the effects of travel. He dealt with the symptoms until February 25 before he finally went to the hospital. He died on March 13.

This sequence of events is now known as the start of the epidemic of severe acute respiratory syndrome (SARS), which infected more than 8,400 people and left just over 800 dead worldwide. Unlike the flu, this virus was completely new and left both medical staff and epidemiologists confounded. The effect of SARS on patients was dramatic. The infections started with an unstoppable cough caused by significant amounts of sputum in the lungs. After one to a few days, the patients were in such respiratory distress that intensive care and forced oxygen ventilation of the lungs were the only options. Treatments with antibiotics proved useless, and antivirals had no effect; the infection seemed resistant to everything thrown at it. The only option was to keep the lungs clear of mucus as best as possible and hope for the best. But the efforts were futile in up to 30 per cent of the initial cases, and the victims died by literally drowning in their own mucus.

Without knowing a cause for the condition or how to prevent its spread, health-care staff put their focus on saving lives without any consideration for their own safety. Unfortunately, that meant coming into contact with the patients' coughed-up and sputtered mucus. Eventually, many of these brave souls, some 20 per cent of all cases, came down with the disease, and many of them died. The most notable casualty was Dr. Carlo Urbani, an infectious-disease specialist who tried to figure out the cause of the disease in Hanoi and was the first to notify the WHO about the potential pandemic.

The world looked to technology to at least identify the source, and in April the answer was found, though many had a hard time believing the news. The culprit wasn't some new virus that needed a new name, nor was it an old enemy like the flu or the plague. At the germ level, this was nothing more than the cause of the common cold: a coronavirus.

The coronavirus is one the most contagious viruses known, and its R_0 is estimated at around four to six, depending on the strain. But it is also killed relatively easily through simple handwashing with soap and water and cleaning surfaces with general detergents. While the impact of infection made the virus appear to be a tiger, from a purely physical perspective, it was nothing more than a kitten. At the logistical level, the reaction was swift. New and profound regulations were set in place to prevent any further losses. Toronto, followed by other major cities in the province of Ontario, undertook unprecedented actions to force hygiene measures. The use of masks and gloves was mandatory regardless of the situation, and travel to work was not permitted if any signs of infection were present; even a cough meant ten days at home. At the hospital, the activities of staff were limited, if not halted altogether, through closures of wards—and, at times, whole hospitals. Patients were subjected to a triage process in which they were asked a series of questions to determine whether any symptoms matched those of the condition. Anyone who answered yes ended up in quarantine to be observed for any worsening of condition. A new order was declared preventing visitors from entering any affected hospitals, and only those with appropriate clearance were allowed into the building. Hospitals were as secure as airports to ensure that when SARS entered the hospital, it could not get back out again. In some cases, hospital volunteers were denied access for close to half a year.

The general public was being advised through the media to wash their hands regularly, cover their coughs and stay home if sick. SARS was beatable, and humans were not going to accept defeat. In Asia, the implementation of strict hygiene measures was easy in hospitals, but harder outside and impossible in remote communities. The government reacted by stopping human movement as much as possible. Schools were closed, and many meeting places were deemed off-limits. Shops and workplaces became deserted as unjustified human activity was deemed unacceptable. A SARS hospital was set up on the outskirts of Beijing; all patients with appropriate symptoms were to be sent there to keep them away from the 20 million inhabitants of the city. Whereas Toronto targeted the virus, China focused on the people.

Both measures worked, and within a month, the number of new cases started to decline. By the end of June, the WHO declared that SARS had been a very close call, but thanks to the discovery of the labile nature of the virus, it could easily be controlled and would soon disappear forever. There was also a clear message to the world that better surveillance and response to an unknown threat were needed. In retrospect, SARS had been fully preventable and should never have spread the way it did, yet the lessons learned were implemented both in the affected countries and worldwide to ensure a safer health-care system.

As for the source of SARS, a report in 2006 suggested the virus originated in civet cats, a specialty delicacy in China, but that was never conclusively proven—the virus was also found in badgers and dogs. The actual jumping point of SARS from animals to humans continues to be a mystery, although it may end up being nothing more than trivial fodder for public health officials who are too busy dealing with other similar threats such as a novel coronavirus found in Saudi Arabia. This SARS-like pathogen—nicknamed the Middle East Respiratory Syndrome (MERS)—first appeared in 2012 and infected dozens in the Middle East and Europe, killing over half of them. National public health organizations are on border alert to prevent this virus entering their countries. For them, source is nothing more than a piece of the germ code that may never need deciphering. All they strive to do is stop the virus from spreading before it becomes the next pandemic and heal those who have been infected before they are lost.

THE RISE, FALL AND RISE OF WEST NILE VIRUS

Although not seen as a typical pandemic pathogen, West Nile virus (WNV) is a model for virus transmission and evolution, with a twist. The virus has been known for over seventy-five years, and its natural hosts are birds, including waterfowl and crows. Global spread of the virus is enabled through migration across land and water. Similar to H5N1 but unlike its flu counterpart, this virus can only reside in blood and cannot be transmitted to humans through close contact. Not that this stops it from spreading; the virus has a shrewd means of getting from one host to the next. It uses an insect vector: the mosquito.

After an insect takes a blood meal from a bird, such as a crow, the virus resides in its body until it is transferred to another animal through a bite. In humans, infection can vary from no symptoms to the onset of a flu-like illness, including fever, chills, headache and nausea. But in a fraction of the cases of infection—mainly involving the elderly and those with health complications—the virus can infect the cells of the nervous system as well as the immune cells and head to the brain, where it causes a form of meningitis—a swelling of the brain due to infection. In the event of this happening, the virus can be lethal. Depending on the geographic area, anywhere from two to fifteen per cent of victims perish.

The virus is known to be endemic to Europe, Africa, Asia and Australia, but until 1999 there had been no sign of it in North America. This changed when an outbreak in New York City led to the deaths of birds and humans alike. Sixty-two people were infected, and seven of them died from neurological complications. The outbreak lasted only a few weeks in the month of August, but that signalled the arrival of a new germy foe. While the news appeared to be all bad, many appreciated the opportunity the arrival of a new pathogen presented. For the first time in modern history, epidemiologists would be able to track the progression of a mosquito-borne emerging pathogen and, if possible, how best to control it. A close eye was kept on the virus to determine whether it would be distributed across the country in the coming years or simply disappear.

Surprisingly, it did both. From the beginning of the millennium to about 2007, the virus was widespread, infecting thousands of people in the United States, Mexico and Canada. But as of 2008, the numbers of infections started to drop. The virus apparently was disappearing and presumably heading towards a North American extinction. Many believed it had lost the ability to overcome the immune systems of birds and humans and was without a proper host; the only possible result was its dying off. But that assumption was wrong. In 2012, a resurgence of the virus led to record numbers of infections and deaths in the southern United States. Somehow, the virus had gone into hiding and, when the time was right, returned with a vengeance.

The reason may have been climatic in nature. For the last decade and a half, the United States and Canada have suffered from drought, but 1999 and 2012 were significantly worse than other years. The lack of rain forced crows and mosquitoes to share the same water sources for drinking and breeding. The crows were left vulnerable to bites and to an incredibly vicious infection that can leave a bird completely drained and with hundreds of billions of viruses in its bloodstream. The dying crow is even more susceptible to bites and will transfer the virus back to the mosquito. The virus becomes rampant in the insect population and soon ends up in humans.

But there is one aspect of WNV infection that could not have been caused by the weather. Even if there were a drought that led to an increase in infection, both crows and humans can develop immunity to the virus. Unlike influenza, which evolves each year, WNV is relatively stable and makes few changes. With each year, the number of infections would decrease and reach a point where only a certain level of infection would be reported, primarily in elderly and immune-suppressed individuals. There should be no reason for a rise in cases. The only logical conclusion is that a combination of factors has worked to develop a quasi-perfect storm. WNV has used the germ code to evolve, and the drought of 2012 has given the newly evolved species the opportunity to amplify in crows.

The best way to demonstrate this isn't through the description of specific nucleotide or amino-acid changes, but by observing a seasonal event in Ottawa. Over the fall and winter months, crows from all over the region congregate in a sparsely forested area of the city close to one of the hospital campuses. At its height in the late 1990s, tens of thousands of birds would flock to this region every evening to caw and crackle the night away. But in the mid-2000s, when West Nile virus was at its most virulent in Canada, the population diminished. The virus can be 100 per cent lethal to crows, and the outbreak decimated their population, although actual numbers were never compiled. In other areas of North America, the rate of mortality varied from 40 per cent to 70 per cent, with the highest being found in Illinois, where over 80 per cent of the infected crows died. Back in Ottawa, dead crows were seen all across the city, and public health officials were quick to mandate that a dead crow should be reported but never touched. While never reaching a zero population, the definite reduction in the crows coming for their nightly gab-fest was dramatic.

Over the next few years, the population started to surge. The surviving crows had developed immunity to the viral strain. Soon, the crow population was back to near normal and the roosting trees once again even blacker than night. In 2012, however, with the new West Nile outbreak, dead crows started showing up yet again. While the number of crows diminished only slightly compared to its first appearance, there was little doubt the virus had used the germ code to evolve and was back. No lab work or complex mathematical equations were needed to prove WNV had a cyclical nature, and while it may seem to disappear, there is always the chance it can make a vicious evolutionary return.

FRIEND TURNED FOE

Normally, *E. coli* is a friendly bacterium that helps us to digest our food and make use of the nutrients from our dietary choices. Yet this germ has the potential to take advantage of the germ code and evolve strains that cause infections with potentially lethal consequences. The first such turncoat activities were identified in the 1940s, when *E. coli* turned out to be the cause of meningitis in children. Researchers started to suspect that the bacterium was involved in other diseases and found it was a cause of urinary tract infections. These infections were rare but treatable, and as such were considered to be important in diagnosis but not worthy of surveillance. But in the 1980s, a new strain emerged that not only caused infection but led to several deaths.

The strain is called O157:H7. Much like the case for flu, the classification refers to two proteins found on the external wall of the organism: the *somatic antigen* (O) and the *flagellar antigen* (H). While the designations may seem complex, their origin refers to German designations for bacteria that either move around on a petri plate, causing a shiny bacterial trail (*Hauch*), or stay in one place and do not give off a sheen (*ohne Hauch*). There are 171 different O proteins and 56 H proteins. O157:H7 was first isolated in 1975 in California, but wasn't recognized for its outbreak potential until seven years later, when a group of twenty-five people from different American states experienced bloody diarrhea from the bacterium. The source was found to be improperly cooked hamburger meat.

Initially, cases were sporadic, but as the 1990s came to a close, the incidence rose and deaths started to occur, primarily in the young and elderly. A new form of pathogenesis was also revealed: infection could cause the kidneys to shut down, a condition known as *haemolytic uremia syndrome* (HUS). This potentially deadly complication sparked surveillance activities to determine whether O157:H7 was becoming a significant threat to human health. Although the infection levels were much lower than other food-borne illnesses such as norovirus and *Salmonella,* there were potentially thousands of cases in the U.S. alone. On a global level, a number of cases in Europe, the Middle East, Japan and Australia were identified. Yet the impact of this strain wasn't truly respected until 2000, when the small town of Walkerton, Ontario, was struck with a massive disease outbreak. From May 8–12 of that year, central Ontario dealt with a storm that caused localized flooding. In Walkerton, about 200 kilometres west-northwest of Toronto, the flooding was significant enough to cause an overflow of soil and manure into the well system, contaminating the water supply. The town was used to such flooding and had a system of disinfection to ensure that the water was safe to drink. However, the pumps feeding the disinfectant chlorine into the water were not functioning, and as a result, pathogen-laden water was sent to the residents. As with the Broad Street pump in London, anyone who drank the water fell ill. More than two thousand people in the community of less than five thousand started to suffer, and eventually seven died. A subsequent government inquiry concluded that the outbreak had been caused by the operators of the water pumping station, who had not maintained the town's disinfectant system.

The Walkerton outbreak was a dreadful reminder that the disinfection of water—a fairly simple process—is of vital importance. Reports of other bacterial outbreaks caused by the processing of meat underlined that importance. Unfortunately, the lessons were not well learned; the bacterium continued to infect a wider distribution of the North American populace in the new millennium.

In September 2006, the Centers for Disease Control and Prevention in Atlanta were flooded with calls from twenty-six states, each reporting an O157:H7 outbreak in their area. The sheer volume of simultaneous reports suggested the outbreaks had to be linked, so investigators looked for a common food source. The food turned out not to be meat, whether fresh or processed. It was spinach—specifically, a batch grown and packaged in California.

While contaminated spinach was known to have the potential to trigger outbreaks, the magnitude of this occurrence was simply too great to ignore. The CDC ran an investigation to identify how O157:H7 ended up in the product and came up with some very painful observations. Manure from both livestock and wild animals had spread through runoff rain into the rivers and then seeped into the groundwater used to irrigate fields. The farmers had no idea they were using unsafe water. The outbreak, then, was an accident, and one that led to 199 infections and three deaths.

The emotionally deflating results of the investigation showed that the real cause of the outbreak was the use of intensified farming where livestock are kept. The only real way to prevent infection was to have a buffer zone between the produce growers and the livestock farms, but this was essentially impossible; water resources were already strained to their limits and neither industry could afford to move to a region with a more sustained water supply. The only other alternative was to tell consumers to cook their spinach and other produce while food safety inspections were ramped up to include water sources in their checklists.

Today, O157:H7 can be found in a number of different areas of the world, although there has yet to be any indication of a coming pandemic. Yet the fear is real; unlike other pandemic sources of infection discussed above, the route of this spread was cultivated fresh food. According to the Food and Agricultural Organization of the United Nations, the importation of food products represents more than a trillion dollars' worth of business worldwide each year. Unfortunately, many countries aren't able to prevent endemic pathogens from being exported through contaminated produce and meat products. Food security measures in many parts of the developed world, including inspections at ports of call and regular testing at government laboratories, have been increased. However, these pathogens will continue to spread in a way not unlike plague centuries earlier—although, instead of rats and fleas, the sources may end up being leafy greens and hamburger meat.

LOOKING TO THE FUTURE

The numerous and complex means by which germs can infect can be somewhat disheartening to anyone trying to prevent infection. The potential for a perfect storm can be found in almost any place on earth and could occur at any time, depending on a number of factors humans simply cannot control without significant will and funding. But as meteorologists have shown in the past, forecasting is not a perfect science, though it can be increasingly helpful over time. With each new outbreak, we learn of yet another path a pandemic can follow. There will inevitably be more novel and emerging pathogens, and each will have its unique means of causing infection and threatening a pandemic. In 2013, two new major threats have emerged, including the aforementioned MERS and a new influenzavirus, H7N9. Both have alarmed public health officials, the public and media. But the key to living with the prospect of a pandemic is no different from hearing news of an approaching storm. You need to calmly follow expert advice based on that information. Likewise, in the case of medical alerts, if things look a little too worrisome, it's best to seek answers from public health authorities, surveillance websites like HealthMap and ProMed, and your friendly neighbourhood microbiologist.

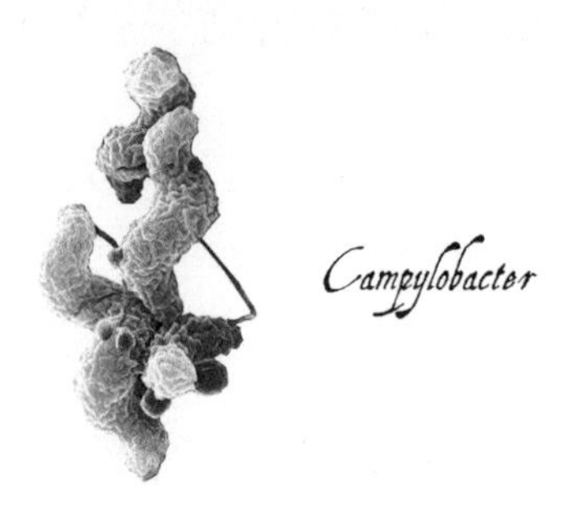

7·

CAMPYING IN THE JUNGLE

In 2001, a microbiologist decided to take some time away from the lab to venture into the wilds of the Amazon Basin in Ecuador. This was, in his view, the trip of a lifetime. The Amazon was not only the most pristine environment in the world, it was also the most diversely populated. He hoped to see and possibly interact with species not found in his native eastern Canada and looked forward to the variety of flora and fauna, the exceedingly warm temperatures and the legendary rains. He was not disappointed; his journey to paradise concluded with a three-hour boat ride away from civilization and into the small village of Mondaña, which bordered the Amazon.

Over the next few days, the researcher-turned-tourist ventured out into the jungle and, with the help of a stunning female biologist guide, experienced the grandeur of the rainforest, integrated with the people, helped out on the farms and even found time to make friends with a local tarantula. At the end of one particularly long hike, he sat down and enjoyed a refreshing meal of chicken, fruit and an exotic juice mélange. He couldn't have been happier.

His joy soon turned to pain, as germs had another plan for him. Within two days, he was more attached to the toilet than to his guide. He immediately started investigating the symptoms: cramps, diarrhea, fever and a constant urge to "go." There were a few possibilities, all of them bacterial. But considering his choice of meal and the fact that he didn't really check to see that it was cooked properly—owing in part to a continued infatuation with his guide—he pretty much knew the cause. He had *Campylobacter*.

The bacterium was first observed back in 1886 by the pediatrician Theodor Escherich—after whom *Escherichia coli* is named—who saw a number of curved bacteria in the intestines of children who had died of food-borne illness. Initially called *Vibrio* because of its active movement under the microscope, the bacterium went through more name changes over the next century than Sean "Puff Daddy, P. Diddy, etc." Combs. Eventually, the taxonomists all agreed to name it after its shape: *Campylobacter* meaning "curved bacteria." There are about forty-five different species, which are usually found in livestock and humans. The pathogen tends to be acquired through the ingestion of improperly cooked chicken meat and takes a few days to start the infection. When it does start, the experience is not quite a living hell, but it can be frustrating.

Similar to other bacterial illnesses, *Campylobacter* causes diarrhea, cramping, nausea, fever and fatigue. It also brings with it the additional annoyance of tenesmus—involuntary spasms of the anal sphincter leading to a continual urge to defecate, and so a nightmarish straining of the entire bowel system, with nothing actually being excreted. The infection lasts between five and eight days and normally requires no treatment.

Our microbiologist continued his journey on the second day of the ghastly symptoms. As he left the jungle, he did his best to hold in what he was sure was trying to get out and hoped he could survive the three-hour boat ride back to town and a toilet.

MEET THE NATIVE GERMS

The case of the microbiologist is not unique. Similar problems strike travellers regularly. All one needs to do is take a stroll into the bathrooms of airports, train stations and bus depots to encounter an aroma of gastrointestinal unhappiness. The air in these places is also filled with germs—respiratory illnesses are continually shed by the unlucky voyagers who happen to have picked up the coughs and colds circulating around the areas they have visited. But unlike the researcher, who actually picked up a pathogen, these health concerns are due in many cases not to pathogens, but to a process more commonly associated with weather.

Acclimatization is normally defined as the adjustment of the body to climatic changes. When faced with a shift in the ambient air temperature, humidity and available daylight, our metabolism, eating habits and circadian rhythms are fine-tuned for survival. For example, a shift from a temperate zone to a tropical one will lead to small but incremental increases in body temperature associated with increased sweating and thirst. A move from sea level to high altitude will result in a change in our breathing—over time, the lungs will expand and gain an increased oxygen capacity; in the meantime, light-headedness, dizziness and nausea may occur. The process is gradual and may require several weeks to find the perfect balance, but equilibrium is eventually reached and the person can function normally in the new environment.

In the context of germs, the same process occurs as the immune system learns to harmonize with the local microbial strains. Yet, unlike the side effects of weather acclimatization, those associated with the immune system's alteration are manifested in symptoms of illness. As the gastrointestinal system becomes used to the new germs associated with food and water, there will inevitably be loose stool and occasional diarrhea. Adaptation in the lungs leads to coughs and mucus production as well as nasal congestion. The skin may encounter rashes, and earaches may occur.

Acclimatization to germs isn't new; even Hippocrates hinted at this phenomenon back in the fourth century BCE in *Airs, Waters and Places*. Because of the (at times stark) differences between two geographic environments, he said, people who travel are more likely to get sick. While he had no scientific evidence to support his claims, he's since been proven correct. Moreover, research in immunology has revealed that this tendency to sickness is part of a natural adaptive process that helps maintain our health. Sadly for tourists, the amount of time they need to adapt to an environment and its endemic germs is far longer than the time spent in a given location.

DON'T DRINK THE WATER

Mexico welcomes tens of millions of tourists each year. People come from all over the world to enjoy the endless beaches, pristine golf courses, historical ruins and lush resorts. The destination does have its risks, including a well-known illness known colloquially as Montezuma's revenge, the symptoms of which are recurrent diarrhea, bloating, cramping and fever.

It's contracted by drinking untreated local water. The aquatic bacterial population is so different from any other place in the world that practically every visitor who comes into contact with it will suffer. The illness usually lasts only about a week and normally resolves without any need for antibiotics. For many travellers who have only a limited time to enjoy themselves, the time required for this acclimatization is far too long.

The obvious way to avoid Montezuma's revenge is to avoid drinking the local water as it comes. But the supply of treated water is limited and can't meet the demand created by the huge number of tourists arriving each year.

In the 1980s, several multinational corporations offered a cost-effective option by taking safe water from various sources—including the municipal supplies of American cities—and bottling it for sale at relatively cheap prices. It was a hit with travellers and is now available in almost every corner of the planet. What was once just a gastrointestinal boon for international travellers is now often consumed at home in preference to perfectly safe tap water.

This might be considered unnecessary. But not, apparently, for domestic travellers.

In 2011, a Canadian television news crew was travelling around various cities and villages in eastern Ontario. One member of the crew was an avid tap-water drinker while another was a bottled-water addict. Over the course of the week-long tour, the reporter on the team noticed that the person drinking tap water had gut issues, while the one drinking bottled water had no such problems at all. The reporter investigated further and found that her two colleagues had eaten the same food, visited the same watering holes and, other than when sleeping, were always together. The only difference was the water source. They had covered an area of less than 500 square kilometres, but the difference between municipal waters seemed to be enough to set off the acclimatization process. On a subsequent trip, the tap-water drinker switched to bottled and was fine.

The simple and easy use of bottled water continues to be an excellent way for travellers to avoid getting sick. However, for those staying longer or moving to a new place, the best solution (provided the local water is actually safe to drink) would be to forgo the bottled water and suffer the week-long symptoms in order to adapt quickly—and save money in the long run.

EAT WELL

Foreign food may offer a cherished gastronomical experience, but there may also be awful consequences. There are almost two hundred food-borne pathogens associated with illness in travellers, and many are common to all regions of the world. Infection can range from a small rumble in the tummy to an outright explosion of diarrhea and vomiting. What is even more problematic is that, unlike water-safety concerns, which can be alleviated by bottled water, the only means of preventing food-related infection is through properly observed safety.

Almost all the countries of the world now have guidelines for the handling, preparation and serving of food. In North America, these have been summarized as the Five Cs: Chill, Clean, prevent Cross-Contamination, and Cook. Properly chilled foods will inhibit bacterial growth. Washing will remove a large percentage of germs from the surfaces of food. The separation of meats and other produce on the preparation counter, and handwashing after each time either is handled, will prevent germs from being transferred from one item to another. And cooking kills germs. The process is simple and can be performed by anyone who has access to a normally furnished kitchen. In many developing countries, however, the guidelines are either unknown or simply ignored.

In the absence of regulated food safety, the burden of preventing illness falls on the tourist, who needs to identify which foods might cause the least amount of problems, whether through acclimatization or infection. There are several basic and easy-to-follow directions that can be adopted by travellers to reduce the potential for disease.

The first is simply to determine whether the place where you intend to dine is clean. Any signs of neglect in this regard may be a sign that the preparation activities are also suspect. Meats, fish and seafood should be properly cooked throughout. Chicken meat must be white, with no signs of pink, while ground meat cannot have any observable redness from blood. Fish and seafood should never be translucent. Side vegetables need to be fully cooked, while raw vegetables have to be washed and sorted to keep out any soil residues. The best fruit is the one you peel yourself. Only use condiments such as mayonnaise and ketchup that are offered in an unopened bottle or individual package.

If you're distracted from giving this level of attention to your food—as was the case with the microbiologist mentioned at the beginning of the chapter, who was too busy socializing to scrutinize—there can be inconvenient consequences. Thank goodness, then, that most food-borne infections are short-lived and can be managed with over-the-counter medications.

A DIP IN THE SCHIST

Eating and drinking are the most common routes to germ acclimatization, although they are by no means the only ones. Swimming, while seemingly inconsequential, can lead to a troublesome infection by the oddly named parasite *Schistosoma* (meaning "split body"). Found in the fresh waters of Africa, South America, the Caribbean and Asia, this pathogen is actually a worm; unlike bacteria and viruses, it is a sexual being and has a multi-stage life cycle. It begins life as an egg left in soils and waters. Under the right conditions of temperature and light, the eggs hatch and release arrowhead-like babies, known as *miracidia*—from the Greek word for youthfulness. These babies seek a particular type of home—namely, a freshwater snail, *Biomphalaria*. Once accommodated, the miracidia grow tails and become wormlike, arrow-headed *cercariae.* These emerge into the water, where they await their next human home. While they can be ingested, most enter the human body just like an arrow by burrowing into the skin and making their way into the bloodstream, where they are taken to the liver. Here, the worms mature. Males are characterized by a middle groove that looks ready to house a spinal column; females live comfortably in the groove, where they mate like bunnies and produce eggs. The eggs go out into the world through the human's urine and feces and the whole process starts anew.

Once in the body, *Schistosoma* first causes swimmer's itch as it starts to dig its way into the skin. If it is not scrubbed off or treated, the next stage of the disease, known as *schistosomiasis,* can occur. In this case, the worms produce progeny not as other worms, but as eggs, which then migrate to the kidneys, liver or other organs. The eggs are destined to be shed through the feces or urine but can inadvertently become entangled in other tissues, potentially leading to such health catastrophes as bladder cancer, genital lesions in women, male infertility, brain damage and heart failure.

While the majority of incidents in both natives and tourists are mild and last only forty-eight hours, there is the likelihood of a form of toxemia known as Katayama fever. In this case, diarrhea is the last of the victim's concerns as the liver and spleen become enlarged, the body starts to break into a severe case of the hives and the skin becomes thick with water, known as edema. The condition can worsen without proper treatment, and although rare, it can be fatal. In the event of Katayama fever, the treatment for *Schistosoma* infection is a single antihelminthic drug, praziquantel. The drug paralyzes the worm and prevents it from living in the blood. When the worm is denied the ability to latch onto a blood vessel, the immune system can easily detect it and break it down into molecules that can easily be excreted in the urine. The treatment is not pleasant, however; potential side effects include dizziness, drowsiness and fatigue. Yet, compared with the debilitating symptoms, the side effects are well worth the trouble.

Humans can acclimatize to the presence of *Schistosoma*, a worm that has been around since the dawn of history. Traces have been found in the mummies of ancient Egyptians who almost certainly died of something else. Though it does sometimes cause serious health problems, the parasite has largely left the human race unharmed. Numerous microbiological and public health studies were performed during the 1980s and 1990s, and a rather interesting picture developed. According to the conclusions of the studies, people who live in areas where the disease is endemic have continual exposure to the worms and may suffer mild infections, allowing the immune system to develop a response and clear the worms. With subsequent infections, the body develops an even stronger defence and eventually forms protection against any further disease. In travellers, however, the joyous experience of swimming in foreign waters may lead to increased exposure to the worms and a greater likelihood of infection.

KEEP THE ITCH AWAY

Mosquitoes are known to carry a number of human pathogens, including malaria, encephalitic viruses and yellow fever, which cause a disease similar to Ebola. But for over a hundred years, a difference has been noted between the effects of mosquito bites on individuals native to a region and visitors. Researchers have been intrigued by the locals' apparent acclimatization to pathogens. Only recently have any of them come up with answers, thanks in part to another mosquito-borne illness called dengue fever.

The origin of the name "dengue" is debated, but it probably derives from *dinga,* the Swahili word for "cramp" or "seizure." The disease has several different stages, starting with chills and a rash that covers most of the body. Next come severe joint and muscle pain, accompanied by bleeding from the nose, gums and sometimes internal organs. These frightening symptoms usually subside after a few weeks, but in a small percentage of cases, they may worsen significantly. Encephalitis is one fairly rare result, as are secondary bacterial infections. Both can be fatal. The illness was first recorded in the 1780s and associated with tropical and warm climates, although there was no indication as to what the reasons might have been. For centuries, researchers tried to find the cause, and in the early twentieth century it was linked to a virus; another half-century would follow before it was recognized as a mosquito-borne illness. Since then, the dengue virus has become a major subject of study in public and global health as the number of cases has grown from fewer than a thousand per year in the 1950s to well over two million cases in 2010.

The dawn of the twenty-first century provided an interesting perspective on the spread of dengue and the apparent difference between those who live in areas where the virus spreads and those who are just visiting. The majority of residents apparently have varying forms of immunity against infection. Moreover, while they still receive mosquito bites, the ramifications are less severe. The nature of this disparity arises from not the behaviour of the humans but of the mosquitoes themselves and how attracted they are to potential prey.

Mosquitoes are drawn to us by chemicals released on the skin. At the molecular level, two chemicals, carbon dioxide and lactic acid, can turn the wings of a mosquito towards its human victim. These two chemicals are released in higher levels during acclimatization to a warmer and more tropical climate, making tourists more delectable to the pests than locals. Thus, the level of infection in tourists is significantly higher. In Thailand, for example, where dengue is endemic, foreigners can have up to a one per cent chance of catching the virus should they stay for up to thirty days. This risk increases should the traveller be in close quarters with many other non-native people. She might, for instance, be part of a tour group, making the odds of being bitten high. However, should there be a larger population of acclimatized individuals, the concentration of attracting chemical is reduced, lowering the chance of becoming a mosquito's meal.

The question then remains: If a traveller is indeed tastier, how can a bite be avoided? There are a number of obvious ways to repel mosquitoes—insecticide sprays, netting, long-sleeved shirts and long pants—but the future may hold a germier answer. There has been much study on whether bacteria on the skin can help reduce the chance of a mosquito bite. The results have been rather intriguing—there has been some association between skin microflora and mosquito attraction.

According to studies out of the Netherlands in 2010, the normal skin bacterium *Staphylococcus epidermidis* increases the level of attraction for mosquitoes; in contrast, *Pseudomonas aeruginosa*—which causes infection in cystic fibrosis patients—appears to ward them off. The team studied the bacterial aroma further and found there were several associated chemicals. This has since opened the door to the manufacture of skin products to prevent bites. So travellers may soon have the opportunity to protect themselves with a squirt of what might be called "Eau de Pseudo."

ANY GERMS TO DECLARE?

The dramatic rise of worldwide dengue cases over the last half-century has been studied, and two significant causes have been identified. The first has been a rise in global trade, allowing infected mosquitoes to hitch a ride on vessels from one location to another. However, this route of spread has been around for centuries without any significant impact on the number of cases detected. The other more important route has been the travellers themselves, who return home with added microbial baggage. Thanks in part to the infected voyagers, there has been a 350 per cent increase in the number of reported cases of dengue in non-endemic regions in just the last decade. The virus has since spread to the local mosquito population and become endemic in new regions of the world, including the United States, Europe and Australia.

As seen with SARS, plague and HIV, travel by people who are infected has become a significant problem and appears to be unstoppable. Legally and ethically, it is almost impossible to stop a person from travelling—and more important, returning home—even if he or she is suspected of being infected.

Not that it hasn't been tried.

In 1995, Canadian authorities detained a twenty-six-year-old man at Lester B. Pearson International Airport for twenty-one days after he told customs officials his mother had died of Ebola earlier in the year. The precaution was not necessary—the man was not infected, nor did he show any signs of infection—but the mere thought of Ebola causing an outbreak was enough to precipitate the action. The attempt at travel restrictions made headlines around the world and led authorities to forgo the procedure in the future.

Six years later, a Congolese woman who had symptoms of Ebola was allowed to enter Canada, even though she was ill and subsequently needed hospital care. While her illness was not confirmed as the fatal virus, suspicion at the hospital led administrators to commence pandemic emergency procedures and left seventy health-care workers extremely worried about their fate.

In 2008, a number of maps of emerging infectious diseases were developed for a variety of pathogens; not surprisingly, less-developed areas of the world presented the highest risk.

These included sub-Saharan Africa, India, Central America and China. However, the results also showed that the risk of an outbreak in a non-endemic area such as North America or Western Europe was closely associated with the presence of a major international airport. While the study's conclusions suggested that higher population densities, as well as higher population-growth patterns, promote the transmissibility and spread of infection, the presence of the airport is key. One particular germ has taken advantage of the air transport system to hamper efforts to eradicate it, and has since become a threat to every human on the planet.

THE MIGRATING MYCOBACTERIUM

Tuberculosis—or TB, for short—has gone by several names over the years, including *phthisis* during the time of Hippocrates, the "White Plague" in the seventeenth century, and "consumption" in the eighteenth and nineteenth centuries. *Mycobacterium tuberculosis* is a pathogen that was first identified by Robert Koch in the 1880s. It looks like a rod—a small rod, of course (hence *Mycobacterium*), and infection causes small, round nodules in the lungs called *tubercules* (hence, tuberculosis). The bacterium is highly stable in the environment and has an uncanny ability to evade the immune system by hiding within a certain kind of immune cell. It can live comfortably in the lungs and spread throughout the body without causing one single symptom. In this latent stage of infection, the body and the bacteria live in relative harmony. But in the event that the immune system is somehow suppressed, *M. tuberculosis* can grow at an incredible rate and spread throughout the lungs and to other areas of the body. The immune system does its best to fight, but cannot stop the bacteria from "consuming" the lungs. Death is inevitable without treatment.

At one time, TB was one of the world's leading killers, taking between 0.5 and 1.0 per cent of the global population each year. However, in 1944, Selman Waksman, a microbiology researcher at Rutgers University in New Jersey, happened upon a fungus with the ability to kill the bacteria in guinea pigs. He conducted clinical trials in humans and found the active ingredient, an antibiotic, was successful at eliminating the disease. The finding led to the mass production and global dissemination of the drug. Within five years, the incidence rate dropped, and by 1980, the number of deaths was less than one in every 100,000 people. There was hope the bacterium could be eradicated altogether. But victory was not to be.

As seen in chapter 4, HIV infection severely weakens the immune system and leaves the afflicted incapable of stopping subsequent infections that lead to AIDS. In the early 2000s, a rash of new cases of TB began to appear in which the infected had little to no ability to maintain latency. The chance of developing consuming TB rose from a low of 5 per cent in a lifetime to 15 per cent *each year*. The incidence of HIV/TB co-infection became so widespread that, in 2004, the World Health Organization created a branch focused solely on this phenomenon. Researchers found that the chances of infection with TB were between twenty and thirty-seven times higher in HIV-infected individuals and that the consequences were far more drastic than they thought as the weapons of battle were eventually rendered useless.

The treatment for TB is quite extensive, requiring a regimen of several different kinds of antibiotics over a period of many months. In addition to the antiviral medications to keep HIV at bay, many co-infected individuals find themselves slaves to the structured regimens of pills specifically timed to ensure that resistance is prevented. Many cannot cope with the routine and simply drop out of programs out of frustration. This lack of completion offers the bacterium the opportunity to develop resistance. And it has.

Multi-drug resistant tuberculosis (MDR-TB) has been a huge problem for health officials for years—but what emerged from the city of Tugela Ferry, South Africa, in 2006 was something else again. "Extensively drug-resistant *M. tuberculosis*"—XDR-TB—was a new strain resistant to almost every possible medication. A study conducted between 2000 and 2004 had identified XDR-TB in the former Soviet Union, as well as Asia and other parts of Africa, but South Africa still took the brunt of the blame.

In 2007, the South African government considered a plan to detain those infected, keep them out of circulation and make sure they received treatment. This ethically dubious scheme was never implemented. But in the same year, the Centers for Disease Control and Prevention succeeded in using detention when it quarantined an Atlanta lawyer who'd been diagnosed with XDR-TB after he had defied medical personnel and travelled abroad. He was treated—forcibly—and eventually allowed to return to society. The case attracted the attention of the media as well as the U.S. Congress, where the patient gave his side of the story. There has yet to be another such detention.

The idea of closing borders to TB-infected individuals, though, appears to be gaining steam. In the early 2000s, several countries took part in a study to determine the impact of migration on domestic TB cases. The rate of infection was shown to be rising, with foreign-born citizens up to ten times more likely to be infected than domestic nationals. Migration was clearly playing a significant role in the spread of the disease.

The response was swift. Many countries implemented new regulations—or enforced existing ones—requiring that individuals who wished to immigrate, including refugees, be screened for tuberculosis. The millions of dollars that could be saved in health care were, it seems, more important than humanitarian concerns.

Although bad news for refugees, the measures have succeeded in their goal of protecting the countries' citizens. The level of tuberculosis in those countries continues to fall. Unfortunately, the future does not look bright for those countries where there are few policy interventions to control the rate of deaths; they could at least double, if not triple, by 2050.

THE ROLE OF TRAVEL MEDICINE

In 1975, a Swiss researcher by the name of Robert Steffen decided to take a look at the risk of infections during travel. He was stunned at the number of cases he observed and published several letters and papers on the high incidence of disease. In 1983, he published a landmark paper in the *Journal of the American Medical Association* in which he reported that, in some years, the attack rate of diarrhea in tourists was over 50 per cent. This startling result led to an even wider-ranging study that yielded similar results. In addition to his searches for illnesses, he found that many travel guidebooks contained little to no accurate scientific evidence to support the tips they offered travellers about maintaining their health. This was not only unacceptable, it was embarrassing.

Steffen spent the next half-decade reaching out to other like-minded professionals, and they eventually held the first-ever conference on travel medicine in 1988. The proceedings were full of firsthand experiences of numerous different diseases, not just diarrhea. After the conference concluded, there was one definite conclusion: a new branch of medical science that focused solely on travel was necessary. In 1991, the International Society of Travel Medicine was born, and while Steffen was not the inaugural president, he was recognized as the pioneer who started it all.

Today, travel medicine has become a major branch of human health, and almost every country in the world has at least one travel clinic. There are now several scientific journals available focusing solely on issues related to the risks of travel and infection. Doctors who specialize in travel medicine are far more aware of the potential problems that can come from travel than family doctors, and many have had experience working in other countries to supplement their knowledge. With a simple visit to a travel doctor, a prospective vacationer can learn not only about the germs that await their arrival, but also how prevalent they are at the destination, how they are spread and what vaccinations or medications are needed to prevent illness.

The level of expertise gained thanks to Steffen's efforts now makes a visit to a local travel doctor not only advisable but paramount. More important, there is equal reason to return to the doctor after the journey has ended. With the potential for a number of unwanted stowaways in the skin, intestines, lungs or blood, a post-trip visit to get the "all clear" is the best means of ensuring that one's health is maintained and the memories of the trip are nothing but joyous.

Back in 2001 in Ecuador, the likelihood of the jungle-faring microbiologist finding a doctor who could help him resolve the *Campylobacter* infection in the jungle was still relatively low. Even with clinics in place, there were few treatment options available. Thankfully, he had made a visit to his local travel doctor a few weeks before the trip to find out what the best options would be for both prophylactic as well as treatment medications. He walked away with a prescription for an antibiotic that would clear any travel-related gastrointestinal infection. As a result of taking this early precaution, he was treating the infection even before he headed back to civilization. On his eventual return home, the travel doctor ensured that he was clear of any gastrointestinal, skin or respiratory infections. As a microbiologist—and, I confess, *the* microbiologist in this story—I can attest that, thanks to the visit to the travel doctor, I brought home only one infection: the "love bug."

8.

THE RAT THAT DOES NOT FEAR THE CAT

One of the tenets of ecology is that predators are feared by their prey, and in the year 2000, a group of researchers from the University of Oxford put that assumption to the test. A rat expert, Dr. Manuel Berdoy; a parasitologist, Dr. Joanne Webster; and a biologist, David Macdonald, acted as nature's mythbusters to demonstrate that nature is not always absolute.

They took thirty-two rats, infected them with a particular pathogen, and then put them in cages where water was placed in one corner and the urine of rats, cats and rabbits in the other three respectively. If nature was stronger than germs, the rats would migrate towards their own odour. But the rats tended to move towards—and even loiter in—the straw that contained cat urine. In contrast, non-infected rats did exactly as nature would prescribe. The discovery of this psychological illness was published, but there was little initial interest in the mainstream media, other than "weird news" filler items about a trio of researchers who had "broken the rules of nature." However, their findings would soon be a critical clue in identifying the cause of a growing psychological problem in humans suffering from HIV infection.

The pathogen used was *Toxoplasma gondii,* a parasite known to cause illness in rats and other rodents. It was first observed in 1908, when Charles Nicolle of the Institut Pasteur autopsied the dead carcasses of rodents called gundis in southern Tunisia. The parasite was unlike anything he had seen before in that it had not one, but three different forms. He found arc-shaped masses of egglike bodies that looked, moved and replicated in a similar way to bacteria. These we now call *tachyzoites,* from the Greek for "speedy life." He found a thick, unmoving, solid cyst now known as a *bradyzoite,* or "slow life." He published his findings soon afterward and gave the parasite the name *Toxoplasma gondii* after the Greek words for arc-shaped life (*Toxoplasma*) and the rodents from which they were isolated (*gondii*).

The natural cycle of *Toxoplasma* infection is limited to cats, rodents and birds. Felines can harbour the parasite without symptoms, and they shed the oocysts through their feces, which are then eaten by rodents and birds. Once inside its new host, the parasite continues to develop more tachyzoites and bradyzoites, which then get reingested by the cat through predation. The ecological cycle is segregated from the rest of the animal kingdom and continues in nature without any threat to human health. But the domestication of these three types of animals allowed *Toxoplasma* to include humans in its life cycle and, thanks to the germ code, develop a mechanism to live and spread in us as well.

Humans become infected by coming into contact with fecal matter or eating contaminated meat. In both cases, the oocysts find a comfortable home in the gastrointestinal tract, where they develop and unleash the tachyzoites—which, like viruses and the tuberculosis bacterium, enter the cells of the body. There, they hide out of sight of the immune system and eventually migrate to the blood, where they can spread. As the number of tachyzoites grows beyond the capacity of the infected cells, they burst free and find other cells to inhabit. Over time, the tachyzoites will infect the liver, bone marrow, lungs, kidneys, eyes and nervous system, including the brain. Eventually, the tachyzoites will form the bradyzoites, which can linger in the body for years.

At the onset of the disease, there are rarely any significant symptoms; the immune system, while stimulated, acts as a balance to allow the continued existence of the parasite without adversely affecting the body—a form of acclimatization. However, in individuals with weakened immune systems, such as those suffering from HIV infection, health complications may occur.

Toxoplasma in AIDS patients can lead to a number of chronic problems, such as ocular and respiratory symptoms, as well as more life-threatening illnesses such as encephalitis. However, some of the afflicted show no clinical symptoms; instead, they exhibit mental ones, including personal detachment, resistance to authority, suspicion and thoughts of suicide. Researchers attempted to prove the association between *Toxoplasma* infection in HIV individuals and symptoms of psychological dysfunction. The findings suggested that everyone who has *Toxoplasma* may be at risk of a psychological disorder later in life, particularly as we age and our immune systems weaken.

From an epidemiological perspective, the information is highly valuable because it offers a better understanding of the origins of psychoses in humans as well as the long-term effects of *Toxoplasma*. From a medical viewpoint, however, the results are merely academic. By the time psychiatric symptoms appear, the parasite has already established itself in the brain. The only options are to treat the psychological disorders and use anti-parasite medications to prevent further migration to areas that control motor function, sensory capacity and memory.

A STUDY IN SEQUELAE

The link between *Toxoplasma* infection and psychological problems is the unusual phenomenon known as *sequelae,* which continues to be one of the most enigmatic aspects of humanity's coexistence with germs. Sequelae are some of the most difficult human health problems to decipher because they usually arise after the normal symptoms of an illness have subsided—and in some cases, after the pathogen has been cleared. Moreover, symptoms are usually associated with chronic illnesses not thought to have been caused by infection.

To study sequelae, then, you need to be a scientific Sherlock Holmes. Whereas those who look at primary infection have the ability to identify and isolate a particular pathogen, a sequela researcher focuses on clues and makes extrapolations from aggregates of information gleaned from a variety of sources in order to arrive at a testable hypothesis. Only then can experimental work be performed. The achieved results can be maddeningly absent, or worse, not statistically significant—the death knell for any theory.

Much like the trio of Berdoy, Webster and Macdonald, several researchers have worked either together or in tandem to hunt down the infectious cause of several chronic diseases. Many have been rewarded with the ability to publish in high-impact journals—a feat akin to winning a gold medal—while others have actually won a golden medal bearing the name of Alfred Nobel. But perhaps most important, these efforts have helped to not only offer possible answers to such infectious lifestyle problems as cancer, obesity and diabetes, but also unveiled possible directions for treatments—and even vaccines—to end their malicious reign over our bodies.

THE WORRISOME WART

Warts were first mentioned in medical texts in about 2500 BCE, but it took a long time before scientists began to pay serious attention to what causes them. That's no doubt because, with the exception of genital warts, their impact on human health (if not appearance) has been minor.

In 1849, researchers in Britain, Germany and France began to see a peculiar phenomenon among poultry workers who were in the habit of inserting a finger into the cloacae of hens to determine whether any eggs were on the way. They all had numerous warts on their fingers. The burgeoning hypothesis was simple enough: warts were caused by an infection linked to fecal matter. As many infections had already been identified through this route, there was little explanation needed. By the turn of the century, other warts—not just those limited to chicken cloacae—were examined and found to be transmissible from one person to another.

In 1949, the virus that causes warts was identified under the microscope, and in 1962 it was classified as a *papillomavirus* (from the Latin word for "swelling"). More than a hundred types of papillomaviruses have since been designated with a number, such as human papillomavirus 1, more commonly referred to as HPV 1.

Papillomaviruses work differently from most viruses, producing few if any symptoms during the first few weeks to months of infection. The virus invades only skin cells. Once there, it tends to play a nasty game of Trojan horse, hiding in the cell and preventing an immune response from being mounted. HPV continues its enigmatic regime by replicating inside the cell, yet not killing it. Instead, the virus takes hold of the cell's natural function and changes it by speeding up the cell's own replication process. Skin cells tend to double in number roughly every four to ten days (which is why we renew our skin layers about once a week), but once the virus takes over, the rate of doubling of the skin cells speeds up to a day or even less. The virus then takes over the physical nature of the cell, forcing it to thicken its walls, making the area harder than the regular skin surface and even more difficult for the immune system to attack. With its fortress intact, the virus leisurely replicates inside the cell, growing in number while the body figures out how to deal with the rapid growth and hardened shell.

The unveiling of the mechanism behind infection was a breakthrough for wart researchers who continued to investigate how HPV had used the germ code to develop such a clever pathogenicity. Their work was valuable to those seeking to understand how viral infections originate and develop and evade the immune system, but in the bigger picture of medicine, there was little interest.

However, the wart world soon gained centre stage in another branch of science. In 1963, just a year after the papillomaviruses were named, cancer researchers came calling and took HPV research on a wild ride that continues to this day.

The word *cancer* was first mentioned by Hippocrates, who described the difference between a simple growth, such as a wart, and a deformation of the skin that could cause severe illness and death. For the next two millennia, the only well-known fact about cancer was that it was incurable; if a malady could be cured, then it was certainly not cancer. In the seventeenth century, with the invention of the microscope, cancer was observed at the cellular level and described as a walled-off section of abnormal cells with the ability to grow at an incredibly fast rate without cessation. Even with this simplistic description, other forms of cancer were soon identified, with almost every internal organ being potentially affected. While the information was helpful in developing statistics, there was still no means of identifying how cancer was initiated or how to stop it, other than through the surgical removal of affected tissue.

By the early 1900s, some of the ways in which cancer grew had been elucidated, including the transformation of cells within an organ to develop a completely different function and the apparent role of environmental triggers such as ultraviolet light or exposure to a range of cancer-causing chemicals known as carcinogens. Yet there were still no clear answers to fully describe how cancer arose.

The first indication that viruses might be associated with cancer came in a series of papers from the laboratory of Dr. Peyton Rous at the Rockefeller Institute for Medical Research. In his experiments, which took place in 1909, a chicken cancer called sarcoma was surgically removed and then filtered to remove any cellular material. He then transferred the liquid to another chicken, which promptly developed cancer. The finding was a major breakthrough: a new and possibly germy cause for cancer had been identified. The research sent many scientists searching for not only the virus—now named Rous sarcoma—but also the mechanism behind the transmissibility of tumours. Unfortunately, while the work was hailed as revolutionary and later was the basis for Dr. Rous being awarded the Nobel Prize, the mechanism by which cancer developed continued to be a mystery.

Another future Nobel laureate, Dr. Renato Dulbecco, identified the means by which viruses can cause cancer. At his laboratories at the California Institute of Technology in 1962, Dulbecco isolated a virus from mice suffering from leukemia and transferred it to mouse cells in a petri dish. It was no surprise to him that the cells became infected. But something else was unexpected: as the mouse cells replicated, they changed, transforming both in appearance and in their rate of growth. Soon, the cells were growing uncontrollably and grouping to form a structure similar to a wart.

The worlds of warts and cancer collided and moved forward together. Research into the mechanisms behind papillomaviruses and several other cancer-causing viruses took off, and within a generation the mechanisms of cancer development were fully understood. The actual molecular functions are highly complex, but can best be summarized as an unfortunate consequence of the virus's Trojan horse takeover and control of the cell. By trying to ensure its survival and propagation, the virus inadvertently sends the cells' reproduction permanently into overdrive. As the virus spreads, so does the cancer, leaving the patient with only two options: either the cancer must be surgically removed, or the patient must undergo treatments, such as chemotherapy and radiation, designed to target and kill cancerous cells.

The association between HPV and cancer led to a hunt for the virus in tumours, although little was learned until a generation later. Over the course of the 1980s, HPV was detected in cancers from all over the body, including the urogenital and perianal regions, the oral cavities, and even the head and neck. Virtually any place where the virus's skin target was present, there was a chance for infection and a risk for cancer.

Amidst all the clinical investigations, one trend emerged. The highest proportion of cases came from those afflicted with HIV infection. Clearly, the virus's ability to transform depended on the immune system. This revelation suggested that the virus was continually transforming normal cells into cancerous ones, but that the immune system was always present to stop the development of a tumour. The news gave researchers the necessary key to pursue a weapon to fight HPV infection even before it could begin.

Over the next fifteen years, the hunt for a suitable vaccine against HPV was on, and a candidate against the most egregious forms of HPV was developed. The targets were HPV 16 and 18, which account for over 80 per cent of all cervical cancers. A second vaccine targeted these two types as well as two others, HPV 6 and 11, which are responsible for genital warts. The vaccines became available in the new millennium, and by the end of its first decade, well over 100 million doses had been administered. Although the current HPV vaccines only stop a small number of cancers, there is significant promise that other cancers soon may be subjected to similar vaccines. There may never be a full eradication of cancer as a result of vaccination, but the impact may one day help to lower the number of cases and possibly displace it as one of the main killers of humans worldwide.

A VIRAL WEIGHT GAIN

Much like cancer, obesity is a major factor in chronic health problems worldwide. Overeating is considered to be the most obvious cause, yet in 1992, Nikhil V. Dhurandhar, a doctoral student in the Department of Food Technology at the University of Bombay in Mumbai, India, investigated another possible trigger for the disease. He believed the cause was not social, but microbiological in nature and sought to test his hypothesis in chickens. He infected the birds with a particular virus, and then watched as they unstoppably gained and stored fat. The results were conclusive enough to gain him his degree as well as to set him on a lifelong vocation to find a cure for obesity.

The viruses in Dhurandhar's studies were from the *Adenoviridae* family. Adenoviruses are tiny pathogens with protruding spikes that make them look like space satellites. They were first isolated in 1953 from the adenoids of individuals fighting the common cold, which is the reason for their name. There are more than a hundred different species of adenovirus, of which fifty-seven are known to infect humans. They are classified quite simply in the order of their discovery, with Adenovirus 1 (or Ad1) being the one found first. In addition to the common cold, the adenoviruses have been found to cause eye and ear infections, pneumonia and gastrointestinal illness. Most adenoviruses are acute; they cause infection and eventually are cleared up, although some strains are able to persist in the body by hiding out in various tissues. One such strain, Ad36, was discovered in 1980 in a six-year-old girl suffering from a gastrointestinal infection. The virus was catalogued and, while considered novel, was not considered to have much value in the overall perspective of infectious diseases.

Five years after completing his doctorate, Dr. Dhurandhar used his time as a postdoctoral fellow to isolate adenovirus from obese individuals. In 2005, he found that the same strain, Ad36, was present in some but not all obese patients. The publication sent an array of researchers on the hunt to either prove or disprove his findings and determine whether or not there was a germy nature to that social disease.

The collection of results revealed a fascinating mechanism behind viral-induced weight gain. Ad36 somehow activated the immune system to trigger the loss of fats in the blood, causing the body to believe it needed to harness and store fats in the form of fat cells, known as *adipocytes.* This effect caused an involuntary hunger for fat and initiated an unstoppable cycle. The result was massive weight gain that could then lead to other chronic problems not associated with the virus. Although there was no doubt that social factors played a role in the development of obesity, Dr. Dhurandhar suggested an infectious link was established and coined the term *infectobesity*.

As only a very small number of obese individuals possess the persistent adenovirus infection, the medical community has been less than eager to adopt the findings. Yet, even with the lack of full approval, research into the management of adenovirus-based obesity has led to improvements in the overall field. In 2012, Dr. Guglielmo Trovato at the University of Catania in Italy looked at a diet developed for those suffering from viral obesity and learned that, while the diet worked for those with the virus, there were improvements across the entire tested group. The diet involved eating an increasing number of vegetables, fruits, nuts, olive oil and fish while reducing the intake of saturated fats, meat and poultry products. The basis of the diet was the modulation of the immune response by keeping unsaturated-fat levels high while increasing the amount of quickly absorbed nutrients.

The diet has been so effective that it could have been called the anti-adenovirus diet, but the regimen already had a more popular name, based on its centuries of use to maintain proper health. It is called the Mediterranean Diet.

A LOSS OF HEART

In the summers of 1947 and 1948, an outbreak of infection occurred in upper New York State that left more than 150 people sickened with a flu-like illness characterized by fever, malaise and weakness. In the small town of Coxsackie, New York, two young boys came down with a severe form of the disease, leaving them paralyzed and unable to function on their own.

Dr. Gilbert Dalldorf of the New York State Department of Health in Albany came to examine the patients and took samples from their feces and cerebrospinal fluid. Back at the lab, he infected mice with those samples and observed similar results. That the disease could be transmitted suggested a germ was at play, and because there were no signs of bacteria under the microscope, the cause had to be a virus. Dalldorf called this virus "coxsackievirus," and within five years he had developed a classification system for its various strains. The first subdivision was based on the ease of isolation and culture in the lab. Group A coxsackieviruses were easy to culture in the lab, while Group B viruses were harder to maintain. He further classified the strains based on numbers—the viruses would thus be called A1 or B1. At the time, Dalldorf had no idea that within a generation the viruses he discovered would be recognized as major global pathogens that contribute in part to an even more heartbreaking disease.

The first instances of heart failure due to coxsackievirus infection in newborns and young children were documented in Africa in the 1950s. The culprits were members of the coxsackievirus B group. Soon, reports came in from Europe and the United States of other B-group infections causing cardiac complications in children, including myocarditis, inflammation of the heart muscle, and pericarditis, which is the inflammation of the pericardium—the muscular sac surrounding the heart. Most of the children simply could not deal with the infection and died. By 1957, reports of similar infections in adults were being recorded, albeit with a higher survival rate. Still, heart damage from both the initial myocarditis and its recurrences left patients with a chronic disease that increased the risk of heart failure.

Much like Ad36, B-group coxsackieviruses, particularly B3, are able to persist in the body and help to trigger the immune system. However, unlike the adenovirus, when coxsackieviruses make an appearance in the heart muscle, the immune system is triggered. The resulting inflammation and scarring occur for anywhere from several days to weeks, but eventually the process stops. Yet the virus is not cleared up, and when it starts to replicate again, the immune system reacts, causing more damage. Without intervention to control the virus, the disease gets steadily worse and the inflammation and scarring inevitably prove fatal. As in the case of many sequelae-causing viruses, there is no cure, and treatment is limited to rest and patience. In more severe cases of health troubles, routine medications to deal with heart disease may be used, but they have no effect on the virus itself.

THE DIABETES DILEMMA

In the second century BCE, the Greek physician Aretaeus of Cappadocia described another autoimmune disease, which he called "diabetes," meaning "to run through," based on the continual passage of urine from afflicted individuals. The nature of the disease remained a mystery for a millennium and a half, until English researchers in the seventeenth century made a breakthrough by the unlikely means of taste. The urine of diabetic patients contained a higher amount of sugar than normal urine and was therefore sweet. This led to a word to be added to the name of the condition, which became formally known as *diabetes mellitus*—"honey-sweet diabetes."

Over the next two hundred years, the anatomical involvement of diabetes was uncovered, and the target of the problem was found to be the pancreas. This organ lies underneath the stomach and is responsible for the production of a number of enzymes and hormones necessary for proper digestion. When the pancreas is removed, diabetes inevitably occurs.

In 1901, Dr. Eugene Opie, a researcher at Johns Hopkins University in Baltimore, took a closer look at the pancreas in autopsies and found that a subset of cells known as islets—so named because they resemble islands scattered throughout the organ—was severely impacted in diabetic patients. He spent the next twenty years trying to identify the reason for this depletion, but came up empty. However, his research did lead to the eventual discovery of insulin some twenty years later by Drs. Frederick Banting and Charles Best, whose work is described later in this book.

Insulin is a hormone produced by islets, and it is responsible for maintaining the proper balance of sugars in the body. Its functions include the signalling of glucose starvation (better known as hunger), the use of glucose in the digestive tract, the maintenance of glucose levels in the blood, and even the use of glucose in each and every cell in the body. Without proper insulin levels, glucose cannot be properly taken in by the body and it is simply left to be secreted in the urine, giving it the characteristic diabetic sweetness.

The mechanism by which islets are destroyed was finally observed by Dr. Shields Warren of the Boston City Hospital in 1925. Using pancreases from children who had died from diabetes, Warren found immune cells where islets should have been. The invasion was clearly a sign that the body's own immune system had led to the destruction of the islets and subsequently cost these children their lives. But researchers still could not find *why* the immune system would perform such an incredulous act against the body.

In 1969, a group of British researchers from the Public Health Laboratory in Epsom, Surrey, decided to investigate the hypothesis that diabetes was in fact a sequela of infection. They took the blood of more than a hundred diabetic patients and searched for coxsackieviruses, already known for their effect on the heart. Every sample had an immune response to the viruses. The study was repeated over the next five years, with all subjects testing positive for a previous coxsackievirus infection. The researchers had every reason to believe the cause was in sight.

Over the next two decades, several other studies confirmed the presence of coxsackieviruses in patients, with B4 the predominant strain identified. At the turn of the millennium, with the aid of computer-assisted analysis, researchers found evidence of similarity between virus proteins and islet-specific molecules. These findings have since led to the development of a diabetes-management treatment consisting solely of a piece of protein called DiaPep277. The protein mimics the coxsackievirus protein involved in diabetes and serves to keep the immune system occupied, which in turn prevents it from attacking the islets. Although the treatment is still in clinical trials, it may be possible that we have a diabetes vaccine in the next twenty years.

Despite all the evidence pointing towards sequelae, there is still no clear acceptance that this mechanism is the true cause of diabetes. While virus involvement is significant, there is still no concrete evidence to show infection actually leads to diabetes. There are numerous, nebulous factors that play a role in the disease, including genetics, environmental stressors and lifestyle, so an actual cause may never be fully accepted. What is known, however, is that the discovery of the viral sequela has helped us to understand the disease, and based on the results of the DiaPep277 work, it may one day cease to be a deadly threat to humanity.

SEQUELAE UNDER SUSPICION

Over the course of this chapter, you may have asked yourself whether infections are *really* related to a number of diseases, including several types of cancers, obesity and diabetes. The answer is, quite simply, that sequelae are not always involved in many of the above-mentioned diseases. There can be no doubt that breast, prostate and colon cancers are hereditary and sequelae are not a factor. Skin and lung cancers are associated with behaviour such as sunbathing and smoking rather than prior infection with a virus. In those suffering from diabetes and obesity, clinical studies that focus on genetics, environment and lifestyle as causes far outweigh those looking at the coxsackie and adenoviruses. The study of chronic diseases will continue to look at social factors to best align a series of habits or activities that could trigger the onset of a disease. However, as seen above, there is room for alternate hypotheses.

While these deviations from the norm may be harder to prove and more difficult to accept, there is little doubt that the gains can be well worth the effort. It's to be hoped that in the future there will be room for those who dare to veer beyond the obvious and devote their time and energy to following clues that might otherwise seem innocuous. As seen in the case of *Toxoplasma* in rats, even the most trifling idea can offer insight into a much larger problem. But more important, the discovery of sequelae can help to answer disease riddles that may never be solved by conventional means—and perhaps also offer opportunities to find and develop cures.

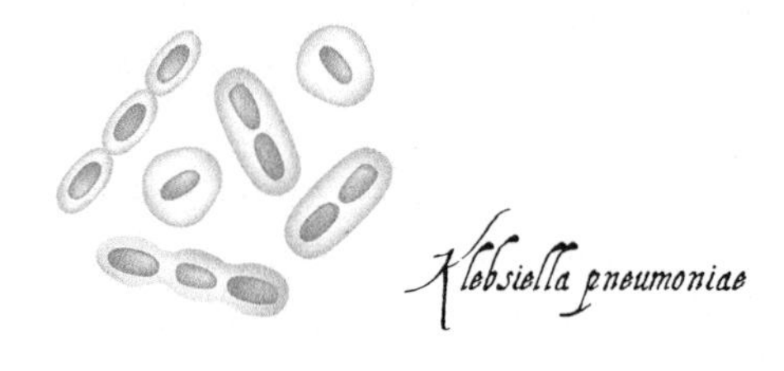

9.

THE GERM CODE STRIKES BACK

Resistance is a continual theme in popular culture, and there is no better example than the *Star Wars* saga, in which an intergalactic evil Empire is pitted against a group of disgruntled civilizations that form an insurgent rebel alliance. The Empire regards these rebels as nothing more than a thorn in the side of the great leader, the emperor. Still, he wishes to crush the alliance. Yet, with the help of a young Luke Skywalker and his uncanny ability to harness a special power known as the Force, the rebels not only thwart the emperor but destroy his greatest weapon, the Death Star—twice—and end the evil reign forever.

This epic has a parallel in the world of germs, although the tables are turned. In this story, humans represent the Empire and germs are small—albeit not hapless—revolutionaries trying to survive our numerous attempts to kill and eradicate them. While they don't have a specific Skywalker equivalent, germs do have their own Force, the germ code, to undermine one of our most powerful weapons, antibiotics. Their insolence has left public health officials reeling as what was once believed to be a panacea in pill form has lost its place as the go-to treatment option. Worse, if germs have their way, within a decade, antibiotics may end up completely useless.

A PREVENTABLE PREDICAMENT

Since Sir Alexander Fleming discovered penicillin in 1928, more than a hundred different antibiotics have been developed. While each has a specific modus operandi, they all perform one of four basic functions.

- *Beta-lactam* antibiotics bind to the outside structure of the bacterial cell and prevent it from reproducing. These include penicillin, ampicillin and vancomycin.
- *Protein production inhibitors* enter the cell and seek out the molecular machinery needed to produce proteins. These include tetracycline and erythromycin.
- *Folate inhibitors* prevent the production of folic acid, which in turn stops the synthesis of genetic material needed for reproduction. These are primarily the antibiotics known as sulfa drugs.
- Finally, there is *ciprofloxacin*—better known colloquially as cipro. This antibiotic targets a specific molecule necessary to allow the genetic material to form proteins. By inhibiting this molecule, the entire function of the cell is halted.

Antibiotics, regardless of the mechanism of action, never work immediately. Rather, they initiate a battle that may last days or months. To ensure victory, a specific concentration needs to be maintained in the body over the combat period to overwhelm the rebels and ensure they are eliminated. This is the reason why an antibiotic prescription states that a pill needs to be taken regularly over a course of days, weeks or even months. If the concentration or duration is lessened for whatever reason, the bacteria have a chance to take a breather and devise a new strategy of defence using the germ code. As history has shown, their reactions are both effective and, at times, deadly.

The occurrence of resistance has been known since antibiotic treatments became available in 1945. In the same year, Dr. Albert Schatz, who discovered a different antibiotic taken not from a fungus but from the soil bacterium *Streptomyces* (Greek for "twisted fungus") and called streptomycin, sent out a warning that certain germs could indeed fight off the drugs and survive. He pointed out that the concentration of the antibiotic was the key to its effectiveness. If the bacterium were allowed to exist in sublethal amounts of the drug, then it could find a way to use the germ code to not only stave off the assault, but also go on the attack and destroy the antibiotic. Over the next five years, other researchers soon joined in the call to use antibiotics properly and to stress to patients the importance of following through with their prescribed regimens.

Sadly, while the research community voiced the risks of improper antibiotic use, the medical community had a difficult time enforcing correct use. Many patients decided to stop treatment early; as they tended to feel better, they would believe the infection to be cleared. This decision, unfortunately, allowed the remaining bacteria to use the germ code to adapt to these sublethal doses. Within a matter of months to years, an antibiotic was rendered useless. By 1950, researchers had to admit that there was little they could do to stop the public from inadvertently helping the defiant pathogens.

But the trouble was nowhere near limited to human misuse. Another industry would take antibiotics off on a path that would lead to the expansion of resistance worldwide. In the 1950s, the agricultural industry viewed these medical treatments as a means of helping protect both livestock and crops from the throes of bacterial illness. The move exponentially magnified the use of these beneficial chemicals; no longer was distribution controlled by the medical community. The widespread dissemination of antibiotics inevitably led to their dispersal and dilution, giving all bacteria the opportunity to adapt and evolve their resistant mechanisms.

The medical community was less than pleased with this turn of events and, with the help of the British government, set up an Agricultural and Medical Research Council Committee to review the practice. When the council's members met in April 1960, physicians and researchers tried to stress that using antibiotics on such a large scale carried significant risk. However, they were outvoted by farmers and members of the agricultural and pharmaceutical industries, who insisted that the economic gains far outweighed the so-called theoretical hazards to animals and humans. What neither group on the council knew at the time was that the risks being debated were already being realized in hospitals, with no chance of reversing the process.

THE BYSTANDER EFFECT

In the late 1950s, resistance to Fleming's penicillin was widely known, and several researchers went on the hunt for other substances to take its place. One of these, methicillin, showed remarkable activity against penicillin-resistant bacteria and was released in 1959. Yet within two years, resistance against this newer drug was already being observed in particular strains of the skin bacterium *Staphylococcus aureus*. This bacterium was first isolated by the Scottish physician Alexander Ogston in 1880 from the abscesses of patients who had suffered from various skin diseases and a blood infection called *septicemia.* Under the microscope, he observed a number of round bacteria assembled in grapelike clusters and called them *Staphylococci,* from the Greek words for a bunch of golden grapes. The *Staphylococci* were subsequently found to play a double-agent role with the body—they were part of the normal microbial community on the skin, but they became invasive once they found their way *into* the body. They were found to cause an array of infections, from the mild but annoying impetigo, in which a wound simply will not heal, to a condition known as toxic shock syndrome, in which the body literally shuts down as a result of infection. But these instances were isolated and normally caused by other health problems, such as a weakened immune system or the introduction of the bacteria into a vulnerable area such as a cut or wound.

When the first cases of resistance to antibiotics were seen in 1945, nobody could work out the cause. The question remained unanswered for nearly twenty years before Dr. Gordon H.G. Davis at the Lagos University Teaching Hospital in Nigeria inadvertently came up with the answer. Between 1962 and 1964, he conducted a study at the hospital to determine the prevalence of antibiotic resistance in patients. He found several strains in infected patients, including *Staphylococcus aureus*, but in these cases, the reason for seeking medical attention wasn't an infection of the skin, but an infection of the sinuses. Davis hypothesized that the *Staphylococci* were just normal flora that did not cause infection, but which, as bystanders, came into contact with the antibiotics and eventually developed resistance.

Over the following decade, the same observation was made in relation to other areas of the body, including the gastrointestinal tract, the respiratory tract and even the eyes and ears. The evidence became painfully clear: colonization of the bacteria did not necessarily result in infection, but still could lead to resistance. Unfortunately, by the time this was learned, the mutinous staph had already developed ways to defeat the medicine on their march to victory.

In 1961, just two years after methicillin was released to the public to control penicillin resistance, a newly resistant strain was isolated from several patients in a suburban London hospital and given the name "methicillin-resistant *Staphylococcus aureus*" (MRSA). While the revelation was disheartening to the researchers who had developed the antibiotic and the physicians who had hoped the new release would be an effective option for treatment, the discovery of yet another resistant strain did not make headlines. But two years later, the first major outbreak of MRSA was recorded at Queen Mary's Hospital for Children in Surrey, England, where nearly fifty wards were contaminated and thirty-seven children were infected despite the use of methicillin. One died. The tragedy sent waves of concern worldwide as health professionals were forced to accept that resistance to methicillin was not only rampant, it could kill. Over the next twenty years, outbreaks in Europe and the United States were recorded, and MRSA was listed as a major pathogen of concern in the health-care setting.

With more resistant *S. aureus* strains isolated over the coming decades, the mechanism behind resistance to penicillin compounds like methicillin was elucidated in the early 1980s. Penicillin antibiotics work by binding to proteins on the surface of the bacterium—known as penicillin-binding proteins, or PBPs—and prevent the bacteria from keeping the external structure intact. In the case of penicillin and methicillin resistance, one or more of these PBPs evolve so that penicillin can no longer bind to them. The bacteria can thus grow happily while the antibiotics are metabolized by the body and flushed away in the urine. The development of this change is entirely due to the germ code, which altered the makeup of the genetic material. With the new protein in place, the bacteria can maintain its presence, continue the infection and ultimately spread to others.

As the prevalence of MRSA in hospitals worldwide increased over the 1970s, many public health officials suggested it might also be spreading in the general community.

That hypothesis was borne out as the first community-acquired MRSA outbreak occurred in 1980, in Detroit. Forty people showed up at the Henry Ford Hospital with varying skin conditions, all demonstrating both *S. aureus* infection and a resistance to methicillin. The majority of cases were linked to intravenous drug use, but a number had no history of those injections. The patients were treated in special isolation conditions in an unsuccessful bid to prevent the spread of the bacterium. In less than a year, the community strains were identified as causes of infection in over 60 per cent of the cases. The information was a clear warning sign that without proper interventions MRSA was going to be a significant problem for the immediate future and beyond.

There was no way to prevent the increased resistance in MRSA to other antibiotics. By the late 1990s, MRSA had developed resistance to nearly every antibiotic in the arsenal, and these germy rebels could be found in every corner of the world, hiding in plain sight. For every thousand normal *S. aureus*, there was at least one resistant strain lurking, waiting to strike. The fight against the bacterium was so ubiquitous that seeking them out and destroying them was no longer an option. The medical community had no choice but to turn to other avenues, including better surveillance of patients and new admissions, reducing the use of antibiotics against MRSA unless absolutely necessary and, more important, spreading the wisdom of proper hygiene to the public. Thanks in part to these measures, the rate of MRSA infections has continually dropped since the turn of the millennium, and in some health-care locations, the incidence of the bacterium has all but disappeared.

STAY A WHILE AND SHARE

News of the discovery of MRSA was limited to scientific journals and conference lectures. Indeed, the discovery of a new mechanism of antibiotic resistance is rarely reported in the mainstream media. In 2010, though, a manifestation of the germ code sent news organizations into a frenzy. The reports focused on a piece of genetic material, a gene called NDM-1 (for New Delhi Metallo-beta-lactamase) that conferred resistance to all but one antibiotic. The gene had first been isolated in 2008 in Sweden from a fifty-nine-year-old Indian patient who had been suffering from an apparently untreatable urinary tract infection. At the time, the case was considered to be an isolated one. However, when several more people in the United States began showing up in clinics with the same symptoms, the true extent of the problem became evident. Not only was the resistance present in other areas of the globe, it was spreading. Within a year, cases of NDM-1 infection were found in Canada, Europe, Africa and the Far East.

The original case was found in a strain of *Klebsiella pneumoniae*, a known pathogen not only of the lungs but also of the urinary and gastrointestinal tract. The bacterium was originally discovered in 1880 by a German researcher named Dr. Edwin Klebs who was looking at the bacterial nature of pneumonia. He found a selection of germs, including what appeared to be long rods. A few years later, in 1882, another researcher, Dr. Carl Friedländer, described the same pathogen in the lungs of patients who died of pneumonia and concluded that these bacteria were the cause of death. The bacterium was named after its discoverer, *Klebsiella,* and the resultant disease, *pneumoniae.*

Over the following decades, *K. pneumoniae* and several other species were found, not only as normal flora in the gastrointestinal tract, but also as causes of a variety of diseases, including urinary tract infections, meningitis, sepsis, liver and kidney failure. The major underlying factor separating flora from disease was a suppressed or weakened immune system; thus, the bacterium was normally relegated to infections in the hospital, where those suffering from a weakened immune system tended to end up. Infections were usually treatable with penicillin without much concern, so infections were regarded as rare and unproblematic.

Yet in 1967, the bacteria began to show resistance against these types of antibiotics, and by 1969 the scientific community had begun to accept the possibility of a newly evolved strain of resistant *Klebsiella*. Unlike MRSA, resistance was manifested through the production of a new worker protein, an enzyme called a beta-lactamase—which, as the name suggests, breaks down beta-lactams. The discovery meant yet another wrench in the cogs of antimicrobial treatment. Still, researchers embraced the opportunity to learn exactly how resistance arose. From a purely scientific perspective, the information was thrilling, but for the cause of public health, the results were altogether demoralizing. *Klebsiella* somehow gained resistance through the acquisition of pieces of genetic material from the outside environment. Once these minor cell monarchs, known as resistance or R factors, made their way into the cell, they would start pumping out lactamases, giving the cell a fighting chance to survive.

The finding was remarkable, but researchers were not content. They believed the R factor could be shared. In 1969, a team from the Medical College of Virginia proved those suspicions to be correct. They "mated" resistant and non-resistant strains to see if resistance could be transferred. In every case, it could. This made the situation even worse for physicians; the presence of one *Klebsiella* bacterium with an R factor could instigate the development of resistance in the entire population. Based on this observation, an alert was sent out to the medical community that resistance could increase through "crosstalk," and that due to this occurrence, not only was the gastrointestinal tract at risk, but so were the lives of patients.

While this crosstalk was shown between like bacteria, a study out of the Soviet Union in 1977 hinted there was more bad news to come. In a paper published in a little-read Soviet journal, a hypothesis was tested that the R factor could also be transferred between different types of bacteria. Few in the medical community paid attention, but in 1989, Dr. Robert Tauxe, a microbiologist at the Centers for Disease Control and Prevention, took the findings to heart when he took a closer look at a strange case of shared antibiotic resistance. A strain of the dysentery bacterium *Shigella* had somehow gained the same resistance as a completely different bacterium, *E. coli*. Tauxe decided to investigate the R factor in each, and found they were exactly the same. Tauxe's paper sparked a series of follow-up studies in hospitals worldwide, and by 2002, the worst-case scenario for public health officials had been realized. All the enterobacteria were found to perform interspecies sharing. The situation was even direr: now the entire germy populace of the gut could potentially turn against antibiotics, rendering them useless and leaving the patient at the germs' mercy.

Epidemiologists tried to identify the behaviours that allowed resistance to arise. But unlike MRSA, hygiene wasn't the determining factor; demographics were. Resistance was associated with underlying chronic health problems—and an aging population with a tendency to spend an increased amount of time in the hospital under antibiotic therapy. As nothing could be done about the aging or those health problems, the focus turned to the use of antibiotics. A program known as antibiotic stewardship was introduced to minimize the prescribing of antibiotics in the hospital. Surveillance of antibiotic resistance became an ongoing practice in many hospitals, if only to keep track of the changes in the germ makeup of the institution. Finally, the identification of any new form of resistance was immediately reported—and, if necessary, sent to the CDC to be recorded and disseminated to the whole medical community.

There was one additional factor missed by the changes in surveillance, and it only became clear after the arrival of NDM-1 in the U.S. Inasmuch as public health officials wanted to keep a focus on local medicine to control resistance, little attention had been paid to the incredibly high rate of antibiotic usage by tourists. With a greater number of travel-related infections, the use of foreign health centres by tourists was also on the rise. There was also a noticeable rise in the practice known as "medical tourism," in which an individual travels to a developing nation to have normally expensive surgery—such as an organ transplant—at a significantly reduced cost. In both cases, there was an increased potential for infection with a novel resistant strain that could then share its resistance with the rest of the still-unaffected native microbial community.

The inevitable happened in 2012, when both Canada and France reported community-acquired cases of NDM-1 containing *Klebsiella*. While there are still a few options available to fight off infection, there is acceptance that the choices are declining and that there soon may be no options to stop NDM-1 from spreading and causing infection.

THE PROBLEM OF PIGGYBACKING

As the plight of antibiotic resistance became more widely known in the 1960s, Ciba Specialty Chemicals developed a new antimicrobial, Irgasan DP 300. The chemical was similar to an antibiotic in that it entered the cell of a bacterium and shut down the process of making the lipids that are necessary for growth and reproduction. At high concentrations, this new chemical was lethal to the bacteria, while in lower doses it hindered reproduction enough to prevent infection. The company took advantage of this latter trait and followed a different path than for antibiotics, selling the chemical, now named triclosan, as an additive to household products to increase their germ-fighting powers. Over the next two decades, the miracle chemical was purchased by many companies specializing in consumer products such as soaps, toothpastes, clothing and even toys. The items were labelled "antibacterial" and marketed as being safer than products that didn't contain the substance.

But in 1984, the veneer began to wear off. Evidence of resistance had been found, and some bacteria even appeared to thrive in triclosan-containing products. A closer look at these insurgents revealed new proteins that not only attached themselves to the triclosan molecule but also escorted it out of the cell, a process known as *efflux*.

By 1999, several more bacteria had demonstrated resistance to triclosan, although the data did not appear to hamper its promotion and use. At the same time, a prominent researcher in disinfection, Dr. Denver Russell at the Welsh School of Pharmacy, became rather worried about the newly described triclosan resistance and postulated that bacteria might be able to harness the adaptation for use against antibiotics. He set to test his hypothesis by investigating whether training *Staphylococcus aureus* to be resistant to triclosan could eventually lead to antibiotic resistance. His experiments, however, proved to be for naught; no triclosan-resistant strains developed that added resistance to antibiotics.

But Russell's efforts sparked another researcher, Dr. Herbert Schweizer of Colorado State University, to take a wider look in 2001 at the possibility of cross-resistance in other bacteria. He chose *Pseudomonas aeruginosa* as his test germ and confirmed Russell's theory: the bacterium gained antibiotic resistance in the presence of triclosan. But the link had less to do with training the bacteria to be resistant to triclosan than with a piggyback effect. The development of triclosan resistance did not increase the likelihood of antibiotic resistance; rather, it helped already-resistant bacteria to prosper during treatment. Schweizer found that the efflux system could perform a dual role, expelling both triclosan *and* a series of antibiotics. Schweizer used this information to look for other possible cross-resistance mechanisms, and found a number of other bacteria—including *Mycobacterium tuberculosis* and *E. coli*—that could perform the same function. As for *Staphylococcus aureus*, there was no such link. Russell had the right idea; he had simply chosen the wrong bacterium. In the following decade, the list of bacteria with the ability to gain resistance to both triclosan and antibiotics grew, suggesting that triclosan was doing more harm than good.

The work of Russell and Schweizer strongly indicates that triclosan use should be decreased or even stopped. And indeed, the fate of the chemical may already be sealed, not as a result of these studies, but of others focusing on an environmental concern. Triclosan has proven to be an ecological toxin that can accumulate in the environment and pose threats to fish and, possibly, humans. In 2012, governments worldwide began to consider whether a partial or full ban was prudent. Whether this signals the end for triclosan will be determined over the coming decade; however, the experience should be a signal to other chemical manufacturers to ensure they take the germ code into effect before they launch another miracle product.

THE CASE OF THE "SUPERBUG"

In 1935, Dr. Ivan Hall and Elizabeth O'Toole at the University of Colorado School of Medicine were engaged in the microbiologically intriguing (but aesthetically disgusting) task of isolating microbes from the fecal matter of newborns when they came across a rather interesting finding: in several of the children, a new bacterium was observed under the microscope. The two worked hard to isolate and culture the strain, although it was no easy task. Eventually, they were able to grow the bacteria in an oxygen-free environment, known as an anaerobic condition. The bacterium was long and slender in appearance and produced small, oval-shaped bodies characteristic of bacterial spores. They named their discovery *Bacillus difficilis* for its rod-like appearance (*Bacillus*) and the difficulty of culturing it (*difficilis*). A few years later, other researchers found the same strain, although it was more like another genus called *Clostridium,* based on the Greek word for "spindle"; the name *Clostridium difficile* was chosen and was accepted by the taxonomists.

For the next few decades, the bacterium was identified as the cause of a number of infections, though they were not gastrointestinal in nature as we know them to be today. Then, the focus was on dermatological and urogenital infections. But *C. difficile*'s role in the gut would eventually be unveiled in 1977, through a collaborative effort between members of the University of Michigan Medical Center and Johns Hopkins University. Several patients on antimicrobial therapy inexplicably developed an incredibly painful disease known as pseudomembranous colitis. The condition was characterized by continual inflammation of the colon and the development of raised yellowish extensions of the gut lining, known as plaques. These plaques are filled with pus and blood and can rupture, leading to bleeding, sepsis and, if left untreated, death. In every case, *C. difficile* was isolated from the plaques and declared to be the cause of this painful illness. While no one died in the study, the number of deaths due to *C. difficile* rose, and by the turn of the millennium the bacterium was recognized as a severe threat to public health.

The study of *C. difficile* grew in popularity, and soon a clearer picture of the process of infection emerged. As Hall and O'Toole suggested, the bacteria are ubiquitous and can be a part of the normal microbial flora in the intestine. However, in the event that the normal fecal flora is altered by the antibiotics, the bacterium can start to develop colonies adjacent to the cellular lining of the colon. As the bacteria grow, they produce a particular toxin that targets the cellular lining and causes the cells to die off and peel away from the rest of the gut. This leads to microscopic breakages in the affected area and the consequent stimulation of the immune system. However, as the bacterium is separated from the cells, the immune system cannot cope; it does, however, stay continually activated, leading to inflammation and even more cell death. The cycle continues without end until either the bacteria are eliminated through the use of antimicrobial treatment or the body dies from the subsequent sepsis.

The study also introduced another complication to the use of antibiotics prior to infection. Looking back at the history of *C. difficile* infections, there was a significant connection with prior antibiotic treatment. This implied not only that antibiotics could in fact be causing harm rather than helping, but that they might become a source of antibiotic resistance. Yet, rather than finding just one mechanism, researchers found many, including the ability to trigger resistance in neighbouring bacteria. *C. difficile* was not just an individual rebel; it was a recruiter as well.

This discovery suggested that antibiotic treatment in the presence of *C. difficile* might be the trigger for gastrointestinal infection; much to the chagrin of researchers, the hypothesis was correct. When antibiotics were used, *C. difficile* began to thrive into numbers high enough to start the process of colitis. When the regimen was stopped, the bacterium's presence decreased, reverting to a form of gut flora. The conclusion was hard to accept, but it could not be denied: with respect to *C. difficile*, the medical community was up against a superbug that could render treatments useless.

Up until the early 1990s, antibiotic-resistant *C. difficile* was solely a health care–related infection. Yet, over the next two decades, the rates of community-acquired infection rose, partly because of the increased use of antibiotics in the home. In 2012, a retrospective on antibiotic-resistant strains was conducted by the world-renowned Mayo Clinic in Rochester, Minnesota, to gain a better understanding of the impact of community-based infections. Their findings were surprising and depressing. Between 1991 and 2005 in one U.S. county, 157 of 385 cases of infection (41 per cent) occurred in the community. Prior to this, infection rates in the community were believed to be less than 10 per cent of the total *C. difficile* cases. Much like in the case of *Klebsiella,* the sharp increase was partly due to the increase in elderly individuals with weakened immune systems who needed antibiotics. But the increase in this particular demographic was much lower than the remarkable upsurge in the number of infections.

There had to be other reasons behind the explosion of cases, although researchers were not entirely sure as to how to explain that finding. Based on a collection of epidemiological studies from 2000 to 2010, no single social factor was discovered to be involved. Instead, almost every aspect of human society was implicated. Resistant strains were found in meat and vegetable products, a consequence of their presence in agricultural sources such as water and livestock. A wide variety of pets carried the bacteria, including dogs, cats, birds and reptiles. Even play areas for children were not safe, as the bacteria at nurseries and daycares were implicated in the spread. No matter where researchers looked, they inevitably found evidence of *C. difficile* and subsequent infection.

The options for treating gastrointestinal illness were becoming increasingly few, so a group of Swedish researchers investigated an alternate direction, using organic treatments rather than chemical ones. In 1983, a team at the Institute of Clinical Bacteriology at Uppsala University in Sweden tried to cure infection using the byproducts of a rather unlikely source: a healthy human intestine.

Over four days in January 1983, a sixty-seven-year-old woman with a recurrent *C. difficile* infection was given enemas filled with the feces of an uninfected person. Within a few days, she was feeling better, and six months later, she was back to normal. When asked if she would go through it again, she was adamant that she would without a second thought.

Despite the rather unappetizing nature of the practice, there was a success rate of close to 75 per cent in resolving the disease—and, in more than half of those cases, the pathogen was cleared up entirely. Subsequent attempts have shown even greater success—better than 90 per cent. Perhaps not surprisingly, more than half of those who underwent the treatment would choose it again should they ever become infected. While the treatment continues to gain interest and acceptance in the medical community, the typical retching response when the topic is brought up outside of institutional walls suggests it's not quite ready for public consumption.

THE FUTURE REQUIRES ALLIANCE

While fecal transplantation may offer hope to those infected with *C. difficile*, other antibiotic-resistant rebels will not so easily be defeated. The list of useful antibiotics continues to dwindle, and the number of new options is slim, with only a handful of antibiotics approved since the turn of the millennium. In order to gain the upper hand, the human Empire needs to refocus its efforts and veer away from a dependence on weapons such as antibiotics and disinfectants and towards an unlikely ally: non-pathogenic germs. Much like the recolonization of the intestines with highly diverse normal fecal matter from a healthy person, humans may benefit from increasing the levels of helpful germs that pose no threat to health.

The idea may seem novel, but it's actually a very old theory stated by one of history's greatest advocates for health, Florence Nightingale. In her treatise *Notes on Nursing,* she wrote of the benefit of opening windows and allowing fresh air into hospital wards. Her aim was to remove the stagnant and uncomfortable air caused by the sick and dying, not to prevent infections acquired in the hospital. Yet, as time wore on, there was a significant fall in the number of life-threatening illnesses. While she didn't know it at the time, she was increasing the microbial diversity of the hospital environment and, in the process, diluting the levels of pathogens.

In 2012, a team of researchers from the University of Oregon followed Nightingale's lead and opened the windows of a hospital ward to see if they could stem the tide of infections simply by increasing the diversity of the germs in the environment. They not only saw the hoped-for increase in diversity, they also proved Nightingale right: fewer infections were noted.

The irony of this experiment is that while we face a possible antibiotic-resistance nightmare, we may indeed overcome revolutionary enemies, not through battle but through dilution. We already know the outcome of waging war with heavy weaponry; it is doomed to become obsolete. Yet, by simply working with the germs that pose us no harm, we may be able to find a way to resolve the rebellion as well as prevent any future insurrections.

10.

GERM SCENE INVESTIGATION

In the investigation of major crimes, the most useful types of forensic evidence are fingerprints and DNA. But according to microbial ecologist Dr. Noah Fierer at the University of Colorado, there is another unique human trait that may be useful in identifying a suspect of interest to the police.

In 2008, Fierer looked at the diversity of bacteria on the skin of the hands and found that each of us has a unique microbial community. With this in mind, he set out in 2010 to test whether these bacteria could be transferred to surfaces such as computer keyboards and then be traced back to an individual. He and his team first swabbed a number of desktops and isolated the genetic material from what they collected. They matched the sequences of the genetic material to their corresponding germs and created a list of microbes—a "print." They performed the same collection from the fingertips of people known to spend time in that area and then attempted to find any matches. The results were quite impressive; Fierer found definitive links between the "prints" from the keyboards and those taken from the employees' fingers. He could easily and accurately identify the specific user of any keyboard.

The experiment was a resounding success and opened the door to a new realm of forensic options. Although this type of "printing" has yet to be used in a criminal trial setting, the potential for yet another means of solving crimes has been set.

Fierer didn't explain the apparent uniqueness of an individual's microbial "print," but a year later, Dr. Rob Dunn, a microbial ecologist at North Carolina State University, came up with an idea that would offer a possible answer. He wanted to identify the body's microbial mass—known as the microbiome—and see if he could find any links between the germy makeup and a person's history or lifestyle. In January and February of 2011, Dr. Dunn headed to two separate scientific conferences armed with sterile cotton-tipped batons and recruited volunteers to swab a portion of the body: the belly button. The decision was not random; the navel is a great place for bacteria to thrive because it is warm, moist and also isolated from the environment. Whatever he found would be directly related to the person and not to their current activities. The volunteers, many of whom were as amused by the experiment as they were intrigued, offered up their navels and endured the slight tickling in the hope of learning more about their invisible colonizers.

Back at the lab, Dunn developed a list of microbes for each person. His results were similar to Fierer's—each person's bacterial collection was unique—but they added a completely unexpected twist to the story. After confirming the presence of a particular germ, Dunn then went in search of its natural home. In several cases, the germs were not local to the community where the person currently lived; rather, they were known to be endemic in other areas of the world, from forests to coastal regions.

Dunn spent more time navel-gazing over the results and eventually came up with a theory that perhaps the microbiome wasn't simply a reflection of the present, but could provide a look back into a person's history. He went back to the participants to find out where they had lived over their lifetimes, and in almost every case, he matched the domiciles of the past with the germs he found in the present. Certain bacteria were associated with regions where the volunteers had been born or raised as children, while others were related to regions where individuals had stayed while attending university or during previous employment. While the data was not always conclusive, Dunn found the belly button to be a microbial museum of lifetime experiences, and through a simple swab, the geographic history of an individual could be traced and revealed.

A MICROBIAL ACCENT

Dunn's results were enthralling, not only to his lab but to the volunteers, the scientific community and the media. Yet he merely confirmed the nature of our continued relationship with non-pathogenic germs. The link between the microbiome and chronic health problems was first hypothesized in the late nineteenth century, around the same time Koch was developing his postulates for germ-related disease. In 1883, the Swiss physician Dr. Hermann Fol, a leader in experimental embryology with expertise in the developmental patterns of chickens, found the relationship between the body and germs mesmerizing. He postulated correctly that microbes came in three different groups, those seen in chapter 3: opportunistic pathogens, mutually beneficial germs (such as those used for fermentation) and those that appeared on and in the body, yet apparently had no impact on the human condition. His words sparked an interesting line of research in the latter case, as microbiologists began to ponder why these microbes existed in the body and whether they actually had a role.

Over the century that followed, a clear picture of the interaction developed, revealing that these germs, while apparently not having a specific function, worked together as a whole to keep the body content. The microbes first act as a complement to the immune system, offering protection against pathogens by overwhelming them so that they cannot cause an infection and are forced to leave the body in search of a more comfortable home. They also work to keep the immune system on alert so that, in the event of a pathogen entering the body, the response is quick and effective. But most of all, they keep the body balanced in a harsh environment that is continually dynamic and always unpredictable.

THE MICROBIOME PROJECT

The United States National Institutes of Health decided in 2007 to spend five years identifying the genetic material of all germs that live in the human gut and to discover any links these had to disease. The group, known as the Human Microbiome Project, was an epic consortium of more than two hundred researchers from eighty universities and scientific institutions across the country. The study focused on close to 4,800 swabs from 242 healthy people. The results of the mass undertaking were released in 2012 and showed that the microbiome is even more complex than imagined. More than ten thousand different species were identified, and no one species was found in all individuals. Moreover, just as Fierer had found in his small investigation, each of the 242 subjects had a completely unique microbial profile. Fierer, though, had simply wanted to match the germy print to a person. Drawing associations between the microbiome print and human health was simply impossible.

While HMP was underway, other microbiome-related projects were also being conducted. In Europe, the MetaHIT project aimed to link the microbiome to chronic diseases, including obesity and inflammatory bowel diseases. Another project sponsored by the Broad Institute in Massachusetts focused on the impact of antibiotics on the development of the infant gut microbiome. Closer to home, research teams such as American Gut and uBiome focused on receiving samples from members of the public and determining their microbiome. In addition, several independent researchers were conducting their own, smaller scale microbiome studies. Taken altogether, the research has provided a comprehensive perspective on how humans and germs relate from the moment of birth until the very last breath.

A LIFELONG BOND

When we are born, we are devoid of microbes, having spent the last nine months inside the sterile environment of the amniotic sac. But the moment we enter the birth canal, we become one with the germs. The first introduction comes from the womb (in cases of natural birth), as well as the skin and saliva of those who touch and handle the infant. Over the coming year, breastfeeding introduces germs from the skin of the mother, while bottle-feeding exposes the child to a number of different bacteria, allowing the microbiome to diversify. The continued introduction of these bacteria is necessary to keep the body balanced, especially if antibiotics are used in the event of an infection. In such cases, the microbiome is significantly afflicted and can even be wiped out in the most severe cases.

For the next decade, the child's microbiome changes both in numbers and diversity as the body is faced with a continual barrage of varying germs. The immune system develops a relationship with compatible harmless germs and learns to work with them to initiate an attack when a pathogen is present. As the body enters puberty, the immune system is well developed and stable, yet the chance to increase the diversity of microbes still exists. Changes in diet, exercise regimes and geographic location—as Dunn found—introduce new types of germs that could eventually find their way into the germy fold.

One of the most impressive changes in the microbiome occurs during interpersonal relationships, when two hearts collide. Germs will inevitably be shared through touching, kissing and other means; if there is harmony, the microbial bond will adapt to remain stable in the presence of the other. But if there is too much of a difference between the germy natures of two potential partners, the immune system will act in defiance, believing the other to be a foreign invader rather than a possible companion. The effects could lead to unanticipated immunological consequences, such as allergies and rashes, as well as psychological problems, including depression and anxiety. From a purely microbiological perspective, the concept of having a successful relationship could depend on being with a person with a similar microbial profile to one's mother.

As the body ventures into middle and old age, the microbiome becomes a major player in health as immunity begins to wane and needs help from its microbial partners to keep pathogens at bay. While infections are more common, the microbiome puts up a blockade against invasion and gives the weakened immune system a fighting chance. As was the case with the sixty-seven-year-old woman in the previous chapter, the mere introduction of healthy microbiome germs was enough to stop the *C. difficile* infection and bring back a more comfortable existence. Even at the end of life, having a strong microbiome is important to prevent any opportunistic pathogens from entering the body and shortening the natural progression towards death. While the microbiome cannot prevent the inevitable, it may help to prevent an early exit due to an infection.

CHRONIC CHANGE MEANS IMMUNOLOGICAL RAGE

While the microbiome is no doubt a friend to the body, there is a caveat: the microbiome is only effective if it is kept relatively stable. At any point in a lifetime, a dramatic change that reduces or eliminates friendly germs and replaces them with unrecognized or even pathogenic ones will trigger the immune system. Unlike an infection, when a full attack occurs, what follows is more of a long-term cold war: an unceasing process of inflammation in which the body no longer watches for infections but assumes that one is imminent. Constant inflammation leads to chronic fevers, fatigue, and pain resulting from localized tissue damage and scarring. The suffering may be manageable with over-the-counter medications such as anti-inflammatories, but more severe problems may require the use of immunity-depressing drugs, which can then predispose the body to other infections or such problems as high blood pressure, osteoporosis and even psychiatric disorders, including depression.

The first studies involving aberrations in the microbiome began shortly after Fol released his postulations on the three different types of microbes. Researchers looked at a number of chronic diseases, including intestinal and bowel diseases, skin complications and even allergies, although at the time the technology was simply not available to decipher any conclusions from the data. The uniqueness of the microbiome rendered researchers perplexed. But over the course of the twentieth century, several revelations were made thanks to the study of two egregious chronic diseases that had been known since the dawn of history. Not only was the necessity of a stable microbiome learned, but also how working with germs could help to stop and possibly cure them.

COLITIS AND CROHN'S

In 1888, the London physician William Hale-White documented a potentially lethal disease in which the colon became inflamed and continually bled, leading to breaks in the gut and the potential for sepsis. He called this "ulcerative colitis" and believed the cause for the disease might have been bacterial in nature. Though he had no means to prove his theory, researchers who followed his lead found a significant reduction both in the number as well as the diversity of microbes in the guts of sufferers. There was also a greater prevalence of one bacterium, *E. coli*, in colitis patients, suggesting a possible link due to a common strain. Yet, other than these findings, there was no smoking gun to fully incriminate this or any other bacterium. The only options for treatment were medications to control the inflammation.

Forty-five years later, in 1932, the American gastroenterologist Burrill Bernard Crohn at Mount Sinai Hospital in New York observed a similar inflammation and ulceration of the intestine, though it wasn't in the colon, it was in the ileum. He also noticed that sufferers had long, tubular growths that appeared similar to those of tuberculosis lungs and felt that whatever "gastrointestinal tuberculosis" bacteria he could find inside would be the cause. Unfortunately, he came up empty. While the disease was named after him, Crohn was left unsatisfied with his work. He had wanted to find a cause and a cure, but just like Hale-White, all he could offer were theories and the inevitability of lifelong medications to control and manage the symptoms.

For the next sixty-five years, there were few advances to help patients of ulcerative colitis and Crohn's disease. The only real success was the isolation of the supposed cause of gastrointestinal tuberculosis in 1963 by Dr. Edith Mankiewicz at the Royal Edward Laurentian Hospital in Montreal, Quebec. As Crohn had suggested, the bacterium was related to tuberculosis, although it wasn't a new discovery. The bacterium turned out to be *Mycobacterium avium subspecies paratuberculosis*. As the name implies, the bacterium was primarily found in birds (*avium*), although it was also found in livestock and was responsible for the development of tuberculosis-like illness (*paratuberculosis*). Mankiewicz suggested a possible link between Crohn's and the ingestion of contaminated meat or milk products, but there was little proof, as non-milk drinkers also turned up with the disease. Even when there was a possible answer, only more questions resulted.

For another fifty years, the situation appeared to be hopeless, but thanks to a new advancement in microbiological research, a major step was achieved in 2010. A consortium of twenty European and Asian research centres aimed to finally identify the problems by taking a big-picture view of the microbiome. They took fecal samples from 124 people who were either healthy or suffering from a number of intestinal problems, including ulcerative colitis, and then, just like

Fierer and Dunn, isolated the genetic material and matched the individuals to their germs. As expected, the fecal matter of colitis and Crohn's patients had far less numbers and diversity of germs than healthy people, but the extent of that difference was completely unexpected. Practically all gut microflora was different in sufferers than in healthy people.

The widespread dissimilarity was shocking; no one had anticipated such a dramatic shift in the overall microbiome composition. In the case of Crohn's disease, the deviation was so great that no similarities between ill and healthy microbiomes were detected. The observation led some researchers to question whether the transformation was unique to an individual or whether it could be transferred to another, spreading disease in the process. Experiments with mice suggested the disease was transmissible, but the evidence was still circumstantial. What was evident was that, for the first time, a charge could be brought against the microbiome in connection with the onset of disease. However, another study was necessary to ensure the argument was sound.

DEVIATING FROM THE CORE

In 2011, the laboratory of Dr. Willem de Vos at the University of Helsinki in Finland took a different approach to investigating the microbiome's contribution to disease. He took several fecal samples from healthy Finnish people over a period of seven weeks and tried to link any microbial changes to medical complications. Much to his surprise, gastrointestinal infection did not lead to a drastic change either in bacterial diversity or numbers. Instead, medical problems happened when non-microbiome bacteria found their way into the gut. If these foreign strains were present, symptoms including diarrhea, nausea, urgent bowel movements and heartburn occurred. When the infiltrators left the body, the problems disappeared.

De Vos then changed his focus to identifying the bacteria that were always associated with health. Of the thousands of different types of microbiome strains, a set of only about 450 bacterial species was present during the healthy times and either reduced or missing during illness. He called this the core of the microbiome. Not surprisingly, when he compared this data with the MetaHIT project, the core was found in only healthy individuals. De Vos concluded that overall health was based on maintaining the core and that any alterations would result in some kind of illness. Unfortunately, de Vos had no idea which of the 450 or so core bacteria could help to resolve issues such as colitis and Crohn's. But he was also unaware that another group of researchers had already narrowed down the number to just a handful, and that they would use these core microbes to solve and even cure these troublesome ailments.

BRING IN THE GOOD BUGS

In 1989, intestinal microecologist Dr. Roy Fuller at Britain's Institute of Food Research postulated that the introduction of a select group of live bacteria in animals could improve the microbial balance of the intestines and lead to a better quality of life. He had conducted a historical review of the bacterial nature of several fermented and soured foods known to play a role in good health. Most contained two types of bacteria: *Lactobacillus*, known for their ability to sour lactose-containing products such as milk; and *Bifidobacterium*, which have a characteristic Y-shape (or split, which in Greek is *bifidus*) and are known to ferment a number of different sugars. Fuller further suggested that the restoration of these two bacteria into the gut could help to protect against infection, reduce the inflammatory condition and possibly cure the diseases.

Over the next twenty years, a number of clinical trials followed, both proving and disproving Fuller's postulations. In almost all cases, colitis was rapidly and effectively managed by these two bacteria, although relapses did occur without continued supplementation. In the case of Crohn's, however, there were few benefits, leaving patients yearning for a solution.

The answer came in 1982, thanks to a serendipitous discovery by Dr. Robert Ducluzeau at the Institut National de la Recherche Agronomique. He was searching through a freezer filled with a collection of bacteria gathered over the previous century and happened upon a particular strain of yeast known as *Saccharomyces boulardii*. The strain had been found back in 1923 by a cholera microbiologist, Dr. Henri Boulard, during a trip to Indochina to learn about an apparently magical tea that prevented cholera. The tea was made with tropical fruits, but Boulard cared less for the wonderful taste than he did the microbial constituents. Culturing the tea, he came across the yeast strain and, after isolation and testing in his lab, found that it was also effective at making wine. Knowing that wine was also helpful in keeping cholera away, he figured he had a bona fide solution that might one day prevent and cure cholera. Seeing the opportunity, he named the strain after its sugar-fermenting properties (*Saccharomyces*) and then himself (*boulardii*).

Sixty years later, in 1982, Ducluzeau was interested not in cholera but in overall gastrointestinal disease and how the strains he found in the freezer could contribute to—or worsen—the impact of infection. He gave the yeast to mice as food supplements and looked for any differences after he infected them with gut pathogens. He tried hundreds of different microbes before *S. boulardii,* yet few ever gave positive results. But when he gave the mice the cholera-busting yeast, the results were shocking. Not only was the number of infections at bay, but their psychological state improved as well. In the next year, two researchers from the general hospital in Celle, Germany, Drs. Klaus Plein and Jürgen Hotz, put *S. boulardii* to the test, giving it to twenty Crohn's patients to determine whether there were any benefits. The results were promising: the symptoms of the disease were lessened and the patients' overall psychological state improved.

Plein and Hotz could only suggest that this yeast could be the answer to Crohn's—their data was simply insufficient to make any real recommendations. The wonders of *S. boulardii* in Crohn's have since been examined in the lab and through clinical trials, and while it is still considered an option rather than a cure, there is little doubt that for people suffering from the disease, there is at least hope for a better life.

GERMS AND WORMS

In 2000, while *Lactobacillus*, *Bifidobacterium* and *S. boulardii* were still under lab and clinical investigation for their benefits in colitis and Crohn's, Drs. David Elliott and Joel Weinstock at the University of Iowa were taking a unique approach to solving the chronic diseases. Rather than looking at the person, they examined the geographic and temporal relationships of these diseases in the overall populations. They uncovered two key clues that offered a possible direction for treatment. The diseases were primarily found in temperate climates, and the rise in cases seemed to match the disappearance of intestinal worms called helminths. These infections are associated with poor hygiene and are common in the developing world. However, improvements in water quality and sanitation gave modern societies the opportunity to prevent infection and illness. As the worms disappeared, the incidence of chronic illnesses rose.

The information, while not conclusive, was too significant to ignore, and in 2003, the two conducted a small clinical trial in which they fed a type of helminth known as a hookworm to seven people, four of whom had Crohn's and three colitis. Six out of the seven patients went into remission that lasted for weeks into months. When the worms were administered regularly over three-week intervals, the symptoms of illness all but disappeared. No adverse effects from the treatments were noted, and every patient was more than happy to continue living without pain with these worms. A year later, the researchers widened the study with similar results: 80 per cent of those treated became symptom-free. Over the coming years, similar studies out of Norway and Australia also proved Elliott and Weinstock to be correct. These researchers had little doubt that the treatments were both effective and safe over the long term. But public health officials were not entirely thrilled with the means of treatment and insisted that a mechanism behind the benefit be elucidated before providing their approval.

As the trials demonstrating the validity of controlled and monitored hookworm infection in the modern world were ongoing, in Brazil, Dr. Jeffrey Bethony was working to achieve the opposite goal—he wanted to eliminate helminth infections altogether. His goal was to develop a vaccine against hookworms to prevent people from the onslaught of anemia and wasting seen in the developing world. He looked at how worms interacted with the immune system and somehow turned off the inflammatory response in the gut.

Bethony took samples from individuals who were infected with great numbers of the worms and analyzed their immune state. What he found must have been demoralizing to him: the worms acted in the gut to send the entire immune system into a state of calm. From his perspective, by relaxing the entire body's defences, the worms could thrive in the intestines without any worry. The fact also meant that a vaccine, which primes the immune system to fight future battles, would be nearly impossible to develop. The worms' use of the germ code all but cancelled out the immune system, even if it was primed.

The discovery was frustrating for Bethony, but good news

for Elliott and Weinstock. They had the answer behind the positive results of treatment; the immune system was finally being given a chance to relax. With more clinical trials, researchers learned how to provide worms in a controlled manner so that they would benefit the patient without any fears of taking over and causing infection. With nothing but excellent results coming out of the lab and clinical trials, there was no doubt that the treatment would be commercialized, and by 2007, several companies began to offer treatment to help solve Crohn's and colitis. The price is still burdensome, costing thousands for every treatment, but for those suffering from the debilitating effects of these two microbiome-mediated diseases, the savings in quality of life are enormous.

A BURGEONING UNDERSTANDING

The world of microbiome research is still young and there will be many more revelations about the links between germs and humans over the coming decade. The entire medical world will be revolutionized by this project. The HMP data already hint at the link between microbial makeup and autoimmune conditions such as eczema and psoriasis. In the case of these two ancient diseases, the skin microbiota may become overwhelmed by a pathogenic strain of *S. aureus* that the weakened immune system simply cannot fight. Clinical trials have led to the development of skin-friendly probiotics to help fight off the bacteria and soothe the discomfort.

As mentioned in chapter 7, the attraction of mosquitoes to the bodies of travellers may also be linked to the microbiome.

Mosquitoes prefer the odour of the skin bacterium *Staphylococcus epidermidis* over that of another normal flora constituent, *Pseudomonas aeruginosa*, so they will be discouraged by a microbiome that had been modified by "Eau de Pseudo" bacteria products and fragrance.

The potential for the future is theoretically endless, as humans learn more about their germy counterparts. Thanks in part to the forensic methods of identifying the good and the bad germs, we may finally learn how to live with them and more importantly, find new and natural paths to better health. We finally have the knowledge that enables us to work *with* microbes to improve our lives.

But perhaps Dr. Peter Singer—a colleague and the current director of Grand Challenges Canada—has a more succinct perspective. According to him, when it comes to health, there's really only one thing we need to know.

It's the microbiome, stupid.

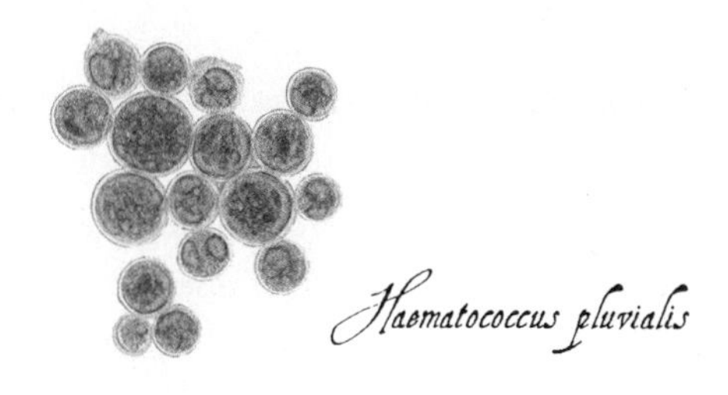

11.

THE GERM ECONOMY

When a researcher makes a discovery, he or she is soon struck by two questions. The first is "Can I get away with attaching my name to this?"—as in the case of Henri Boulard's *Saccharomyces boulardii* or Theodor Escherich's *Escherichia coli*. And then there's the matter of how the information is going to be sent out into the world.

The usual route is to publish a detailed account of the experimental achievements in an appropriate scientific journal. Back in the late seventeenth century, Antoni van Leeuwenhoek had only one option for his letters and drawings: he sent them to the Royal Society in London, and they published his works in their annual journal, *Philosophical Transactions*. Today, researchers look to prominent—and preferably open-access—scientific journals to publish significant findings such as the discovery of a new pathogen, while other less important observations, such as the topographical observation that Kansas is flatter than a pancake, are targeted to more specific publications, such as *The Journal of Irreproducible*

Results. But in the latter half of the twentieth century, a third question began to emerge, although the intent was significantly more selfish: "How can I make money from this?"

The economic benefit of research has become an important part of any funding proposal. In 2013, the President of the prestigious National Research Council in Ottawa, Canada made it explicitly clear: "Scientific discovery is not valuable unless it has commercial value." While basic research is primarily limited to the laboratory, every scientist is encouraged to develop a scenario for the eventual commercialization of their work in order to benefit themselves and their sponsoring institutions. In the microbial world, there are prime examples of how germs have been exploited to benefit consumers; you've probably got quite a few at home. Yogurts, kefir, sour cream and cheeses are milk products left to "spoil" in the presence of germs to make a tasty treat. Fermented foods such as sauerkraut and kimchi are nothing more than perfectly "rotted" vegetables. Even chocolate requires the intervention of bacteria in order to transform its original bitterness into a more palatable form. The impact of microbes in the food industry has become so important that multinational food companies now have their own microbiology departments, which focus solely on isolating and culturing the various bacteria and yeasts to ensure that their products have the perfect organoleptic properties (taste, smell, texture and visual aesthetics) for public consumption.

The pioneering efforts of researchers over the last century have helped to improve our well-being and our quality of life. Antibiotics are produced by various fungi and bacteria;

when we ingest probiotics, they help to keep the gut happy. These are the better-known examples of germs at work on our behalf; there are numerous other uses for germs that ultimately benefit us. Thanks to the work of a handful of researchers turned corporate executives, germs have gained a new place in our society in which they are viewed not simply as threats, but also as allies.

TO KILL A KILLER

In 2002, an outbreak of food poisoning swept across the northeastern United States, causing seven deaths. An investigation by the Centers for Disease Control and Prevention revealed that the victims had all eaten turkey products from a single food-processing plant in Pennsylvania and that the infectious agent was *Listeria monocytogenes.* This bacterium was discovered in 1924 by Dr. Everitt Murray at McGill University in Montreal after rabbits being bred for laboratory experiments began dying in large numbers. When Murray took a look at the animals' blood under the microscope, he noticed high levels of bacteria and immune cells known as *monocytes,* which were also infected with bacteria. There was little doubt that the animals had died as a result of a massive, yet fruitless fight against the pathogen, which evaded the immune system by hiding inside the body's own immune cells. He decided to call the organism *Bacterium monocytogenes*, which in 1940 was changed by a rather haughty Scottish microbiologist and self-professed taxonomist named Dr. James Hunter Harvey Pirie to *Listeria monocytogenes* in honour of the pioneer of disinfection and antisepsis—and fellow Scot—Dr. Joseph Lister.

Listeria infection in healthy individuals can result in twenty-four hours of nausea, vomiting and diarrhea. However, in people with compromised immune systems—such as the elderly, the very young, those with HIV infection, and pregnant women—the condition known as listeriosis can occur. Similar to what Murray found in rabbits, the bacterium invades the cells and begins to track its way into the rest of the body. Initially, flu-like symptoms are present, but over weeks to months, as the bacteria make their way to the bloodstream, there can be more dire consequences. Shaking and convulsing are common, and other neurological symptoms such as paralysis may occur. In pregnant women, spontaneous abortions are likely as the bacterium kills the fetus. If the infection is left untreated or treated too late, organ failure follows, and, not too soon after, death.

Listeria monocytogenes is a hardy organism and is found all over the world, primarily in agriculture. It has been isolated in fields as well as the rumens of livestock and poultry, although in the latter there is no infection. The bacteria tend to enjoy warm, moist environments but can easily live in cold and dry conditions. Not surprisingly, *Listeria* can find its way into the food supply without much trouble if sanitation levels are low. In the 2002 case, the sources of contamination were poorly cleaned drains, which aerosolized the bacteria straight onto the food products. Until the investigation was complete, no one knew how the pathogen spread. In hindsight, the outbreak was completely preventable with proper and regular cleaning. But owing to the rapid pace of meat processing to meet demand, there was simply no time in the schedule to ensure proper safety; an outbreak was inevitable.

As more *Listeria* outbreaks began to emerge, public health officials called for better management of food safety through slower rates of production and stronger attention to hygiene. Yet consumerism all but made these recommendations unfeasible. The conundrum prompted the United States Food and Drug Administration to look for options other than simple cleaning and disinfection. Hidden in the scientific literature, they found a possible but controversial option: a means of disinfection that was not chemical in nature, but biological—using another germ, a virus of *Listeria* called a *bacteriophage,* to keep food safe.

Bacteriophages appear round or, in some cases, like microscopic lunar modules. Phages, as they are more commonly known in the microbiology world, are "viruses of bacteria" and have a unique appetite for their germy counterparts. They attach to their victim, send their genetic material into the cell and then multiply, causing the bacterium to expand and eventually burst. Phages are also finicky in their choice of victim and usually only target one strain of bacteria.

They were first discovered by Dr. Félix d'Herelle at the Institut Pasteur in Paris in 1910 while he was investigating an outbreak of dysentery in Mexico. Through the microscope, he noticed the remains of several bacteria in the fecal matter and wondered whether they had been killed by the immune system or another entity. He cultured some of the fecal matter to grow bacteria and filtered off the solids from another portion. He then placed the clear fecal liquid on the cultured bacteria and waited a few days. Much to his enjoyment, he noticed that, in several places on the cultures, there were holes, or plaques, in the bacteria. The remains of the bacteria looked exactly the same as what he had earlier found in the fecal samples. While he wasn't exactly sure of the cause, he declared that there was an entity causing the bacteria's deaths, which he called bacteriophages, or "eaters of bacteria."

For the next few years, d'Herelle's searches found a number of other bacteriophages possessing the ability to kill almost any bacterium, including *S. aureus* and *E. coli*. He tried to sell phage preparations as cures for skin infections in the 1940s, but the appeal of a germ that killed germs was lost on the public. Moreover, with the release of antibiotics, there was little to no appetite for an alternate means of treatment. His products disappeared from shelves, but his laboratory continued to operate in France, eventually being purchased by a multinational, L'Oréal.

By 1998, the world had changed significantly as antibiotic resistance had become a crisis and the extent of food-borne infections had even the World Health Organization concerned about the future safety of food. At the University of Maryland in Baltimore, longtime bacteriophage researcher Dr. Alexander Sulakvelidze believed the time was right for a phage comeback. He came together with a group of physicians and a corporate lawyer to form a tiny company, Intralytix, with a mission to use phages to improve safety in the food industry. The work focused on several bacteria, although their major success came in developing a treatment against *Listeria*.

The process was simple, requiring only the spraying of products with a solution containing a number of different *Listeria*-specific phages. The technique was so effective that, in 2006, the FDA approved the use of the product to prevent contamination of meat products. The company began to sell the technology in 2009. Though acceptance has been slow, the use of phages continues to grow and expand, with new products designed to kill *E. coli* O157:H7 and even *Salmonella*. While there may be no means to fully ensure a safe food supply, the potential presented by phages offers hope that our worries will be lessened and we can enjoy our meaty meals without overt concern.

ASTAXANTHIN AND THE ANGRY ALGAE

One of the weapons the immune system uses to keep the body safe is oxygen. Oxygen is a needy element, requiring attachment to other elements to stay content. Each molecule of breathable oxygen consists of two oxygen atoms joined together (O_2), while water molecules feature an oxygen atom bonded to two atoms of hydrogen (the famous H_2O). If oxygen is left alone—that is, in the elemental state—it becomes highly reactive, searching for and latching onto whatever might happen to be around. When an oxygen atom comes near a cell, whether microbial or human, the outside surface is literally attacked as the oxygen forms bonds with molecules that are otherwise happy on their own. The additional oxygen disrupts the molecule and forces it to change its structure and, in many cases, its function. This process, better known as *oxidation,* can cause regional damage to a cell in low concentrations while higher levels of individual oxygen atoms may create an incredibly hostile environment, leaving the cell with no option but to suffer and die.

During an immune response, such as inflammation, elemental oxygen is produced and allowed to attack whatever might happen to be in the area. While this is an efficient way to ward off an infection, there is collateral damage to the body's cells. If inflammation is maintained over a long period of time, the accumulation of damage can lead to chronic health problems, including colitis, Crohn's disease, atherosclerosis, Alzheimer's disease, Parkinson's disease, early aging and cancer. In the middle part of the twentieth century, several researchers went on the hunt for possible natural remedies to the effects of oxidation and found a viable option in an already-known molecule called tocopheryl acetate, better known as vitamin E. In the lab, vitamin E could reduce the levels of oxidative damage by giving the elemental oxygen a happy partner to form bonds with. When given to humans, there was a significant reduction in the levels of oxidative damage as well as a calming of the immune system during inflammation. By the 1980s, vitamin E was a star in the natural health world, and soon other researchers went on the hunt for other "antioxidants," as well as means by which to produce them en masse.

Shortly after the vitamin E boom, the precursor of vitamin A—known as beta-carotene—was given the spotlight as it demonstrated in the lab and in animals the potential to prevent the onset of cancer. Further studies on the molecule revealed it could do so much more: it could easily handle reactive oxygen species, as it was able to help lower the effect of inflammation as well as spread throughout the body. As the 1990s began, beta-carotene was described as a wonder drug and was being sold as a supplement that could not only prevent cancer, but also help to improve overall body function and even help to reduce the effects of aging. The overwhelming demand sent companies searching for additional sources. While it is naturally found in carrots and red palm oil, the process of its manufacture was costly and burdensome. This led other companies, sparked by researchers, to look for more cost-effective and environmentally sustainable sources. One such company found success in 1999 when it figured out how to turn red algae into gold.

The Hawaiian company Cyanotech, headed by algal researcher Dr. Gerald Cysewski, started in the 1980s in the hope of mass-producing a spectacularly red-coloured beta-carotene supplement known as astaxanthin. The chemical is found in crustaceans such as lobsters, giving them their characteristic colour, as well as in a number of blue-green algae, particularly the species *Haematococcus pluvialis*. *H. pluvialis* was first seen in 1844 by the German scientist Julius von Flotow, who described a blue-green algae about 50 millionths of a metre in size that was common to rainwater-collection areas such as birdbaths (*pluvialis*). What made the algae so different from others was its ability to turn blood red (*Haematococcus*) in the lab. Using a combination of sunlight and salty water, he could make the algae produce the substance, which he then collected, analyzed and found to be astaxanthin.

While Flotow had no idea of the commercial possibilities, the entrepreneurial Cysewski certainly did. He set up shop in Hawaii and put together a giant algae-production facility complete with a small laboratory and giant pools of contained seawater. Drawing from the work of Flotow and others, Cysewski simply put the algae into the seawater under sunlight, sat back and watched the water turn to "blood." After years of perfecting the production of astaxanthin, Cyanotech released a product called BioAstin with the claim that it was one of the world's most powerful antioxidants. The FDA soon agreed, approving the supplement in 1999.

Since then, astaxanthin has gained prominence in the natural-food sector and continues to be tested for other benefits. So far, lab studies have shown that the red chemical may be effective in helping to support joint health, increasing energy levels and even helping to prevent another sun-related problem, sunburn.

FATTY ACIDS WITHOUT THE FISH

Children are told incessantly that they need to eat well to grow up strong, but they tend to refuse foods that taste strange or bad to them. Fish oil, for one. The product has been a staple in many households because it has been extolled throughout history as being "good for you." Over the last century, the functional ingredients in fish oil have been identified and given names such as omega-3 fatty acids, omega-6 fatty acids and docosahexaenoic acid, better known as DHA. The consumption of these chemicals, collectively known as essential fatty acids, has been proven to lead to better heart health with reduced damage due to inflammation, and even better brain function. For kids, however, nothing is worth that taste.

As if to save the palates of children worldwide, the company Martek Biosciences developed a new fish-free product containing primarily DHA and several other essential fatty acids. The product, life'sDHA, was approved for sale by the FDA in 2001, and soon the company began to sell the product, highlighting the fact that it was not only free of a fishy taste, but also completely environmentally sustainable because it came from a microbial plant source: algae. The life'sDHA product sold well and was soon bought by the multinational vitamin producer DSM, which put the product into a number of vitamin supplements, snacks, juices and even cereals.

While Martek's taste transformation provided a grand entrance into the marketplace, the origins of its product were significantly more humble. In the 1960s, there was a general interest in learning more about the nature of algae, which were distinct from plants, animals and bacteria. One focus of the research was the identification of their fatty acids, and several labs spent years isolating and purifying these molecules. During these investigations, a major discovery was the similarity between products formed by the algae and those found in fish oil. There were hints that algae could be used to mass-produce these compounds, but the means to scale up production were simply not available at the time. Moreover, because there were few concerns then over the sustainability of fish stocks, there was little to no interest in taking on such a venture.

That changed in the 1980s, when stocks of fish such as cod were suddenly depleted and fishing quotas were introduced in many countries. The threat to the fish-oil supply was evident, and there was a resurgence of interest in alternate means of mass production without harming a single fish.

Two such entrepreneurial researchers were Dr. David J. Kyle and Dr. Paul Behrens at the University of Maryland. In 1985, they started Martek with the hope of making DHA using sustainable algal means. Within four years they had a grant from the United States National Institute of Diabetes and Digestive and Kidney Diseases to grow DHA from different strains of algae and test them for their efficacy and safety. The grant led to millions more dollars in venture capital, and soon Martek was ready to get to work. In 1994, the first product, an additive for baby formula, was sold in Europe and eventually was licensed worldwide.

But Kyle and Behrens were not done. They wanted to sell their oils as a separate entity, partly to increase awareness of the benefits of DHA, but also to increase sales of the fatty acid. They developed the name life'sDHA and sought out FDA approval for DHA's use in the United States. When it was obtained in 2001, Kyle and Behrens became superstars of the essential-oil world. Soon their product was being sold for several food product lines, and by 2010 life'sDHA was in the majority of products claiming to be supplemented with DHA. With unprecedented interest in the effect of DHA on maintaining a healthy lifestyle, there appears to be little end in sight to their success.

THE GENESIS OF GENENG

While researchers in the algal world focused on the applied use of naturally derived products, bacteriologists were learning how to control the function of germs by playing with their genetic material. Genetic engineering began in earnest in the 1970s, when researchers realized that modifications at a bacterium's genetic level had an impact on its way of life. Production of proteins could be halted by removing fragments of the genetic material; similarly, new products could be made by introducing new pieces of genetic material. The most favourable aspect of these processes was the lack of any significant harm to the organism. Whether genetic material was deleted or added, the bacteria lived and grew normally. The discoveries led to the widespread belief that bacteria could be turned into microscopic factories making a variety of molecules, such as designer antibiotics and even human medicines. However, no one at the time could have conceived of the impact genetic engineering would have on the world in just a few short decades.

During the infancy of genetic engineering, which some in the field call "GenEng," there was only one available bacterium: *E. coli*. Though it was known as a pathogen, a plethora of non-pathogenic strains existed, and many were tested as subjects of genetic manipulation. One particular strain, known as K-12 from its designation in the Stanford University *E. coli* collection, was easy to control, grew to significantly high numbers in a relatively short period of time and, more important, didn't cause infection even when ingested. The strain made for a suitable subject for experimentation in petri dishes as well as in students new to pipetting by mouth—trust me, K-12 doesn't taste entirely bad. K-12 became the standard for testing and was used as the basis for the first commercial genetic-engineering product released to the public: insulin.

In 1921, Canadian diabetes researcher Dr. Frederick Banting, along with his student Dr. Charles Best, began a series of experiments in which they surgically removed the pancreases of canines and then tried to find a way to prevent the onset of the condition. One of their experiments was based on the previously mentioned achievements of Dr. Opie in 1901, in which he identified the islets responsible for diabetes. Banting and Best took islets from the extracted pancreases of healthy dogs, froze and then thawed them to a semi-frozen state in order to extract the juices, and then injected that juice to the diabetic dog. Almost miraculously, the dog returned to normal, although the effect was temporary—the dog needed multiple daily injections to stay reasonably healthy. Banting tried the same experiments on cattle and had the same results. There had to be something in the extracts contributing to the management of diabetes.

The molecule, which had evaded Opie's investigations, took years to find, but eventually Banting and Best were able to isolate it. They initially called it "isletin," but eventually it became known as insulin. Finding it to be a protein, Banting and Best felt there was every indication that the molecule would produce similar results in a human patient. In 1922, they gave the purified insulin to a young diabetic boy, who quickly regained his appetite and strength. The insulin had worked as predicted.

In 1923, Banting won the Nobel Prize, which he shared with Best, and the production of insulin began in earnest. At the time, the number of diabetic patients was reasonably low; the use of purified insulin from cadaveric sources was sufficient. But as the world population grew rapidly after the Second World War, there was a concern that supplies could no longer meet the increasing demand. There had to be a quick and cost-effective way to mass-produce insulin. By the late 1970s, the genetic material associated with the formation of insulin had been identified and the specific gene that coded for the protein was unveiled.

In 1979, a group of researchers out of a new company called Genentech tried to determine whether the sequence could be put into *E. coli* K-12 and mass-produced. They put the insulin gene into another piece of genetic material—similar to the R factor responsible for the spread of antibiotic resistance—called a *plasmid* (from the Greek "to create a connection"). Although plasmids are only a fraction of the size of the genetic material of a bacterium, they are able to function just like the main genetic material, leading to the formation of cellular components that change the way the cell lives. When the plasmid containing the insulin gene was inserted into the bacteria, no one really knew what would happen. But the bacteria not only produced insulin, they mass-produced it beyond expectations. Over the course of just one day, enough insulin had been formed to meet the needs of between ten and twenty people.

The results were so dramatic that Genentech sought out a pharmaceutical partner, Eli Lilly, and in 1982 they developed their bacteria-based product, Humulin. Regulatory approval to sell the product in the United States and the United Kingdom—a process that can take anywhere from two to five years—was completed in a matter of months. On October 30 of that year, Humulin was given full approval for use in diabetic patients. By Christmas, orders for the drug exceeded all expectations. Today, genetically engineered insulin is an integral part of diabetes management and is used by millions worldwide.

THE GERM BOOM CONTINUES

In the early twenty-first century, microbes are improving lives everywhere. Take environmental sustainability, one of the most important aspects of environmental science today. Several newly developed bacteria have been tasked with breaking down plastic waste, oil spills and leftover pollutants from mines and other contaminated soils. Lawn preservation companies have begun to eschew chemicals to remediate lawns beleaguered by grubs. Today, they choose to treat the soil with a microscopic worm called a nematode, which finds grubs incredibly tasty. Even the lavatory industry has jumped on the bandwagon, selling popular bacterial additives for septic systems to help kill pathogens and reduce the intense odour.

Germs may one day be the key to great wealth. In the last decade, a few bacterial species have been found to be effective gold diggers, creating microscopic gold nuggets. While these measure a mere few trillionths of an ounce, the potential for a new gold rush is apparent—instead of picks and pans, the tools of choice will be petri dishes and culture flasks. The same transformation could occur for black gold, better known as oil. Over the last twenty years, researchers have been finding ways to replace the derrick with biological fermenters. There are already several companies producing biofuels using nothing more than bacteria and carbon dioxide. One project being undertaken to achieve the cleanest form of horsepower comes from a group out of the University of California, Santa Barbara. These researchers, led by Dr. Michelle O'Malley, have found a fungus that can create hydrogen from simple sugars. While this may not be surprising—all microbes can use simple sugars—Dr. O'Malley's source of fungus certainly is: horse feces.

Despite all of the incredible work being conducted, the most amazing advancement in the use of germs involves not algae, bacteria or fungi, but viruses. In Ottawa, Dr. John Bell has opened up a new avenue to treat one of the world's oldest diseases, cancer. Bell and his colleagues have developed a number of different genetically engineered strains of cancer-killing germs called *oncolytic viruses*. These viruses work in the laboratory as well as in people. His most successful product is known quite simply as JX-594, the first virus to be used in the bloodstream to target and kill cancer cells. For the aptly named Jennerex Inc., the company helping Bell to develop the product, their pioneering work offers a unique opportunity to launch the most potentially lucrative use of germs in history. For Bell, however, his work will hopefully remove the stigma of germs as solely our enemies and reveal them as potential allies who may one day help us achieve one of the most elusive medical achievements known to man and woman: curing cancer.

In today's world, we don't dwell on the impact of microbes as we imbibe our alcohol, pop our supplements and eat our phage-treated meat products. Yet, without the contribution of these germs, many of our favourite products and valuable medicines might not be in stores or in our kitchen cupboards and medicine cabinets. The manipulation of bacteria, yeasts, algae and viruses for our benefit will continue for generations to come, and there is little doubt both our lives and the condition of the environment will benefit. Despite all that is known about the harmful effects of germs, an equal amount of consideration should be given to just how important they are in keeping us happy and healthy.

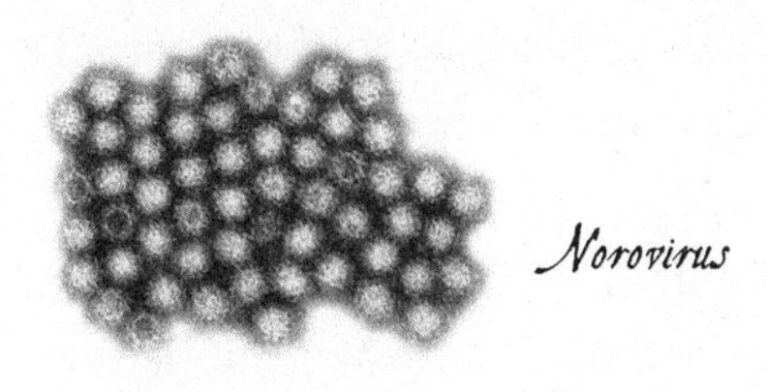

Norovirus

12.

DON'T PANIC

The words appear on the cover of Douglas Adams's immortal invention, that vast reference work known as *The Hitchhiker's Guide to the Galaxy:* DON'T PANIC. While useful as advice in coping with almost all aspects of our existence, these words apply very well to our coexistence with germs.

Of the millions of species, most are harmless, while only a fraction are *mostly* harmless (thank you again, Mr. Adams) and a relatively small 1,450 or so have pathogenic potential. Historically, human beings have tended to focus on the germs that can kill us (and, admittedly, this book contains a lot about pathogens). There's a good reason: it's hard not to be panicky when otherwise healthy people start to get sick and die. Centuries and millennia ago, this reaction was completely justified. The public had little, if any, faith in the ability of their medical and political leaders to stop the deadly attacks. Times of plague, cholera and smallpox left millions in a state of terror, wondering whether they might be the next to succumb to the disease. Yet even today, with our widespread scientific and medical advancements, the public tends to react emotionally whenever an outbreak or worse occurs.

Despite today's wealth of public information on microbiology, epidemiology and advanced medical treatment, as soon as reports appear of SARS or "swine flu" infections, the level of anxiety rises and can reach the point of hysteria.

WHY WE PANIC

The nature of panic in relation to germs is an under-researched field. However, reaction to natural disasters is much better understood, and a germ outbreak can certainly be one of those. One of the pioneers of this area of research is the American sociologist Enrico Quarantelli. In 1954, he released his preliminary results, based on five years of study, on what initiates panic among the masses. He found three specific requirements, including:

- *the inability to identify the threat;*
- *collective powerlessness to stop the threat;*
- *individual isolation leaving one to fend for his or her self.*

With epidemics in the past, a general sense of powerlessness was a natural reaction to an unknown or poorly identified enemy that caused sickness and death. To make matters worse, the lack of any substantially effective medical treatment would leave individuals to fend for themselves, or to passively accept the inevitable. In modern times, however, there is really no need for panic.

When SARS came to Toronto, people who were not hospitalized were free to come and go as they pleased. The hospitals were the only places where the virus spread, and only those who were kept in quarantine were truly in isolation. In the case of the "swine flu," the pathogen was known, the means of avoiding infection were continually shared with the public, and there was a worldwide effort to control the spread and treat the sick. By all accounts, these two outbreaks should have been viewed as problematic, but not reasons for panic. Yet each time, the public went into a frenzy and, encouraged by a fearmongering media, panic spread.

The disconnection between the realities and the eventual public response had little to do with Quarantelli's criteria, but boiled down to another more ingrained aspect of human behaviour: trust. While public health officials and politicians strived to maintain calm and insisted they had a handle on the situation, the public apparently wasn't receptive to their words and instead believed there was more to the story. This attitude was not only misplaced, it was toxic, leading to unneeded fear and concern. The solution to this problem isn't a simple one, however, as it requires people to develop their own sense of trust. And that trust should not be placed in the talking heads on television promoting calm, but rather in the situation at hand. In essence, the key to preventing the panic triggered by SARS, flu and other pathogens is the development of trust in germs and our relationship with them.

THE RISE OF THE GERMEVANGELISTS

Seeing this requirement, a number of researchers have come out of the laboratory and shared their knowledge with the public. In the last decade, this new generation of microbiologists has worked to increase the knowledge of germs by popularizing the findings in labs. The work of Dr. Rob Dunn was a significant move forward in understanding the microbiome, and his use of the belly button as the sampling choice certainly captured many an imagination. The team of Drs. Manuel Berdoy, Joanne Webster and David Macdonald took their observations of the effect of *Toxoplasma* on the rat to the public and gained a great response to their own work as well as to research focusing on the psychological effects of *Toxoplasma* on HIV patients. A researcher at the University of Ottawa named Dr. Robert Smith? (yes, the question mark is part of his name) has taken the spread of knowledge to new heights, transforming his mathematical modelling of the spread of infectious disease into the intensely popular world of zombies, equating pathogens with the undead fictional monsters.

These and a select number of other researchers have become more than germ extollers, they have become "germevangelists." Their work, much like their televangelist counterparts, aims to increase the public's trust in—and adherence to the commandments of—coexistence with germs and to make the public familiar enough with them that they can be discussed in any environment. Moreover, while their work may seem fun and at times outrageous, they are, without a doubt, driven by the intent to tackle Quarantelli's three criteria for panic and ensure that knowledge is gained during the good times and that calm, rather than panic, is the norm during times of outbreak.

WHERE THE GERMS ARE

A mild-mannered microbiologist at the University of Arizona, Dr. Charles Gerba, has spent most of his career focusing on germs in the environment—including tap water, seawater and sewage—and the role they play in human health. In 1999, he added a new position to his resumé: germevangelist. His goal was to help the public ease its fear of the germy unknown.

To achieve this, Gerba armed himself with cotton swabs and petri dishes and spent the next four years sampling more than a thousand different public surfaces in order to determine where bacteria lurk and the risk they pose. The results were a gift to the media: it would be safer to eat off a toilet seat, said Dr. Gerba, than off the average kitchen counter.

A closer look at the data revealed that household items normally taken for granted, such as sponges and dishcloths, refrigerator handles, sinks and cutting boards, were the most likely to contain germs, while the toilet seat—especially if the lid is put down while flushing—had the least amount of bacteria in the home. The office was no better, with phones, keyboards, computer mice and microwave oven handles posing the greatest threat to health.

Gerba set off to find other germy places. He went to schools and found that desks, backpacks, water fountains and cafeteria trays were rife with germs, many of which could cause infection. In the shopping mall, carts were continually loaded with bacteria—three-quarters of them contained fecal bacteria. Worse, money and even recyclable bags were highly contaminated with pathogens. Restaurants were also on Gerba's swab list, and he found that almost everything that touches a tabletop was massively contaminated with germs ranging from environmental to fecal. The tabletop was essentially the germy hub.

Overall, Gerba's work, while at times exploited by the tabloids, was a major effort towards keeping us all calm. He informed us about where germs can be found and which ones pose a threat. He also offered the opportunity for other researchers to head outside the laboratory and share the germy way of life with the rest of the world. As the digital age developed, several researchers uncovered more risks associated with such personal items as cell phones and tablets, service kiosks like ATMs and ticket booths, and schedule maps. No matter what might have been tested, the chance for infection was present.

A few consistent points emerged:

- *unless sterile, nothing is ever free of germs;*
- *high-traffic areas will be the germiest;*
- *surfaces most often touched will always have the most germs;*
- *sharing objects means sharing germs;*
- *where there's raw or uncooked food and water, there will be germs.*

Today, these tenets may seem trivial, yet without the work of Gerba and others, they would be kept out of the public purview and, more important, not considered during normal daily activities. While the release of germy findings might lead to greater caution, the return of such prudence can be quite beneficial. Based on economic studies conducted on just one infection—the flu—simple hygiene may save billions of dollars in terms of decreased absenteeism at work, lower health-care costs and higher retail sales.

CAST OUT THE EVIL SPECIES

In addition to knowing how to spot places where germs are likely to exist, knowing how to get rid of the evil ones is equally important and helps to prevent Quarantelli's second criteria of powerlessness. Centuries have been spent looking at how germs live and die in the presence of natural and chemical factors; this work has led to the development of a set of recommended actions to prevent infection, collectively known as hygiene. Hygiene is the use of a few select practices to overwhelm microbes and either remove them from places of risk or kill them outright. Many of these measures are similar to regular cleaning, but, as germevangelists and other public health officials continually profess, the differences are critical to preventing harm.

Cleaning—the use of soap or detergents and water to remove dirt, grime and oil from a surface or body—does not necessarily lead to safety, microbiologically speaking. All one needs to do is look at the standard cleaning cloth to best appreciate the difference. When used in combination with water and detergent, the cloth can easily make a surface appear clean, whether that is a dish or a tabletop. Yet inside the fabric, germs don't just reside, they grow to incredible numbers. Over time, the cloth will change from an instrument of cleanliness to one of germ spread. To use a dishcloth hygienically, one needs to take measures that are easy to perform, yet may be omitted owing to a lack of either knowledge or diligence.

The first of these is the use of heat. Other than the thermophiles mentioned in chapter 2, germs are vulnerable to temperatures above 70 degrees Celsius; they simply cannot deal with the change in the environment, so they die. In laboratories and hospitals, researchers and technicians have for decades used instruments known as autoclaves to bring objects to a temperature of 121 degrees Celsius (in the presence of steam pressurized to 15 pounds per square inch above atmospheric pressure) to render anything inside sterile. The autoclave has since been adopted in other environments—for example, tattoo parlours, where anything destined to come into contact with a person is treated to ensure that no germs are shared with the client.

At home, autoclaves are quite unnecessary, as there is no need for sterility—the body and the immune system can handle most invasions. Still, we can make use of both heat and steam as part of our daily routines to remove potential threats. Bringing food and water to 71 degrees Celsius converts germs from enemy structures into additional protein, lipids and genetic material for the body to digest. On surfaces, the use of steam cleaners can easily render an area—including clothing—safe. In the past decade, the use of steam has been growing, and is being considered for use in health-care facilities and other industrial sectors. For the cleaning cloth, laundering in hot water will do the trick, as will immersion in a steam stream.

Where heat cannot be feasibly used, or where proper equipment is not available, the practice of disinfection can help to keep an area safe. Disinfectants differ from soaps and detergents in that they are designed specifically to kill germs. All disinfectants have an active ingredient that is responsible for the killing. In the most commonly known disinfectant, bleach, the active ingredient is sodium hypochlorite, which is a combination of deadly oxygen and the even deadlier chlorine. When put together, the active ingredient seeks out anything organic and obliterates it, causing it to go white—thus its use as a whitening agent in many popular industrial processes, as well as in platinum blonde hair colouring. The chemical is both an excellent disinfectant and a welcome addition to other cleaning agents, such as detergents.

Other active ingredients have been discovered from a variety of sources, including coal tar, pine oil, seaweeds, oranges and the tea tree. They have all been tested in the lab and can kill germs successfully while leaving behind a pleasant scent. In the twenty-first century, the majority of these active ingredients are made in the laboratory and the scent is added artificially, but the effect on germs is still the same. Yet for many people, nothing smells more like "clean" than the residual chlorine that comes from bleach. In the case of the cleaning cloth, that smell is a clear indication that a disinfectant is being used.

Both heat and disinfection are great for surfaces, but the effects on human skin may be rather problematic. As anyone who has suffered a steam burn or touched undiluted bleach can appreciate, these germ-killers are also unfriendly to humans as well. The issue of keeping the skin safe—also known as antisepsis—is the third hygienic practice, and quite possibly the most important. In this case, hygiene does not aim to rid the skin of germs—that would be impossible. Instead, the goal is to remove excess levels of microbiome bacteria—and their accompanying smell, which we call body odour—and remove what are known as transient germs. Almost every pathogen is transient, usually making temporary stops on or in humans before heading back into the environment in search of a new home. Pathogens usually tend to reside on the hands after touching a contaminated surface, and are either inhaled or ingested as a result of continual touching of the face. The best way to prevent these transient germs from making a home in the body is to keep those hands clean!

Hand hygiene is a religion in its own right, and it has a select group of simple, easy to perform commandments:

- *Wash hands with soap and water—preferably warm to hot—lathering for about twenty seconds, or enough time to say the alphabet twice.*
- *Washing should be performed a minimum of eight to twelve times a day, and not just after using the toilet. Any time there is a change of location, such as from work to home, handwashing should be performed to gain a fresh start.*

- *Hands should be washed after touching anything that is considered to be high touch, such as a cafeteria tray, a menu or even money.*
- *If the availability of water for handwashing is sparse, then the use of an alcohol-based hand sanitizer for twenty seconds will offer protection between washings.*

The scientific literature is full of research proving time and again that hygienic practice works in the developed world. Yet the health benefits of hand hygiene are epitomized by a group of people who live a less lush life. A group of researchers at Aga Khan University in Karachi, Pakistan, ventured into one of the many slum areas where hygiene was considered to be impossible and tried to reduce the level of infection in the squatter population. Through the use of only soap and water—either dirty or clean—levels of infection went down by more than three-quarters; if there had been a hand sanitizer present, the levels might have been eliminated completely. Considering the impact of soap and water in a completely impoverished environment, the authors suggested that there was no better way to stay safe than to perform hand hygiene. By extrapolation, there should be little reason for any infection to spread in the developed world.

WORKING TOGETHER

Quarantelli's third requirement for panic, "individual isolation leaving one to fend for his or her self," is epitomized by the incidence of hospital-acquired infections (HAI). Patients who are sick, have weakened immune systems and suffer from other complications are vulnerable to the onslaught of infection, which in addition to impairing their healing may take their lives. The need for microbial safety is therefore at its greatest in this environment. Yet patients continually feel they are left on their own when it comes to fighting off germs.

In order to combat this problem, numerous organizations focused solely on patient safety have been formed in the hope of encouraging health-care staff to be more aware of the risks and prevent them from happening.

One of the primary targets of patient-safety advocates was the fact that cleaning and hygiene in the hospital were themselves segregated. Because of their widespread need for hygiene, hospitals had divided the duties amongst two groups: infection control, which focuses on staff and patients; and environmental cleaning, which focuses on the infrastructure. While the two were supposed to work together, there was instead a rivalry, especially whenever an outbreak occurred. The infection control specialists would blame the surfaces of the infrastructure, including bed rails, curtains, sinks and other places where germs were known to hide. In return, the environmental cleaning staff would blame poor hand hygiene by staff, patients and visitors. Both sides had a point: the scientific literature is rife with examples of how a lack of one or the other can lead to an outbreak. But patients began to feel isolated from any real help.

In 2010, a group of advocates, companies and a few willing hospitals in the United States worked to remedy the problem by first calling a truce between the warring sides and then working together to reduce the levels of infection and increase the trust of patients. The team used a multimodal approach that focused on maintaining personal cleanliness and a pristine environment as general objectives rather than to focus solely on hands and surfaces. The efforts were more successful than imagined: hospital-acquired infections didn't just decrease; in some cases, they disappeared. Best of all, patients no longer felt like isolated victims, their perceived quality of care increased and, not surprisingly, their stay times decreased.

Today, the multimodal model is becoming the norm in hospitals and has been hailed by the U.S. government, which now offers monetary rewards to institutions that effectively reduce the level of infection to zero.

The successes are not limited to hospitals. Most commercial establishments have introduced regulations regarding staff handwashing and have implemented recurrent cleaning regimens. From food establishments to airports, the importance of preventing the spread of germs has become a part of everyday activity. Schools are also adopting the model in the hope of keeping children safe incorporating both hand hygiene and cleaning lessons in their curricula. The correlation between more frequent handwashing and cleaning and reduced absenteeism has already been proven in the scientific literature, giving credence to the hygienic actions. Even in the home, many of the personal encounters I've had with those working on hand and environmental hygiene together have claimed lower levels of infection and, best of all, a greater sense of well-being. They also claim to feel less isolated from the rest of the world and have either reduced or lost their fear of germs entirely. While they understand that the multimodal approach cannot prevent infection outright, it can definitely help to stop the spread of disease in their home and, hopefully when the next outbreak occurs, curb their anxiety as well.

COEXISTENCE OVER COMBAT

Up until just a few centuries ago, no one even knew what germs were or what they could do to harm us. But thanks to scientific and medical advancement, our knowledge has increased, as have our chances for coexistence. Gone are the days of polio and plague, while modern struggles such as HIV and chronic infections with MRSA and *C. difficile* are slowly being understood. We appreciate the need for clean water and safe food—even as isolated outbreaks continue—and the influence of proper hygiene as a solution to over-reliance on antibiotics continues to grow. We eradicated smallpox through vaccination, and we aim to stop other relentless infectious diseases in a similar manner. We continue to learn more about our own germy friends in the microbiome and how these good germs can be used as allies to help overcome the threat of aggressive pathogens. We have even learned ways to harness the germ code to improve our health.

There is still a long road ahead before humans and germs live in total harmony. The work of dedicated researchers and public health officials will continue to foster the belief in a future in which outbreaks no longer happen and the word *pandemic* is relegated to history. But they cannot succeed alone; each one of us has to realize that we also play a role.

To that end, this book has tried to offer some insight into some of the actions needed to keep us on the right track. The various stories may have focused on the germs themselves, but each also describes how behaviour has contributed to the onset of outbreaks as well as to their eventual end. Admittedly, the scope the chapters have covered has been vast, and in some cases outside the realm of your day-to-day activities. Most of you may never be concerned with *Schistosoma,* HIV or the brain-eating *Naegleria*. But the activities involved, such as travel, interpersonal relationships and swimming, are part of your daily life, and other, less apocalyptic germs are waiting.

While there may be little chance of infection from these extreme entities, a host of other germs not included in this book may haunt you; of the 1,450 or so pathogens, only a handful was described. While we may not need to worry about these particular germs, we should still heed the lessons they offer, as there is no doubt that others lie ready to take advantage of our relationship.

Personally, I know this all too well. Even with my knowledge of germs and behaviour, I still get sick about twice a year. But rather than worry or panic about the situation, I simply get my rest, take my medications, keep away from others whom I could infect, and just keep repeating to myself: "Don't panic. After all, we've deciphered the germ code. All we have to do now is learn how to apply it to make our relationship the best it can be."

ACKNOWLEDGEMENTS

This book would never have been written without the support of the team at CTV Ottawa, who turned me into the Germ Guy and helped me stray from the lab into the public eye. I am especially grateful to Kimothy Walker, Leanne Cusack and Caroline Ives for always encouraging me to spread the germy word.

I owe thanks to the trio of Dr. Syed Sattar, Dr. Earl Brown and Dr. Lionel Filion at the University of Ottawa. They not only advised me to keep the science solid but also listened to my sometimes-outrageous ideas for making research comprehensible to the public. They also understood that I needed time away from the lab to write the drafts of this book and were gracious in making sure grant proposals, papers and other research activities were kept to a minimum. I owe a special gratitude to Nobel Prize winner Dr. Kary Mullis, who wrote the entertaining *Dancing Naked in the Mind Field* and with whom I've had the opportunity to speak. He has been a continual inspiration and I value his words of support.

I am indebted to Scott Sellers, associate publisher at Random House of Canada, who encouraged me from the time we met to take the leap of faith and pitch the idea for this book. His steadfast enthusiasm kept me confident throughout the process.

I have five very special people to thank for their help during the writing of this book. My mother, Patricia Tetro, who shares the same enthusiasm for learning and sharing, always ensured that my words were not only informative, but also engaging. My father, Peter Tetro, a published poet, helped to keep the words flowing and always warned me when my language and tone would "lose the reader." My friend and colleague, Jason Gilbert, who once promised me "You'll Never Walk Alone," made sure I never took myself too seriously. My great friend Nathalie Earl, who has been a confidante for the last quarter-century, kept me sane with our cherished coffee chats. Her stories of family life with her husband, Chris, and her two children, Madeleine and Zachary, kept me balanced and focused on the people for whom I was writing this book. And finally, my love, Dr. Anastassia Voronova, who has been with me and cheered me on since the debut of the Germ Guy, always found the time from her own busy schedule to read every single word of every single draft, and made sure that my essence and love of life never disappeared from the text.

Most of all, I have to thank my editor, Tim Rostron, who kept me on my toes without ever uttering a negative word, and who in every returned draft offered me a perspective that would make me smile. I gained so much from his insight and experience as I learned how to take what I've learned in the lab into the real world.

To all of you, my deepest, sincerest, germiest thanks!

INDEX